# Crop Management: Towards Higher Yield and Sustenance

# Crop Management: Towards Higher Yield and Sustenance

Edited by **Corey Aiken**

New York

Published by Callisto Reference,
106 Park Avenue, Suite 200,
New York, NY 10016, USA
www.callistoreference.com

**Crop Management: Towards Higher Yield and Sustenance**
Edited by Corey Aiken



International Standard Book Number: 978-1-63239-131-5 (Hardback)



Printed in the United States of America.

# Contents

# Preface

The world is advancing at a fast pace like never before. Therefore, the need is to keep up with the latest developments. This book was an idea that came to fruition when the specialists in the area realized the need to coordinate together and document essential themes in the subject. That's when I was requested to be the editor. Editing this book has been an honour as it brings together diverse authors researching on different streams of the field. The book collates essential materials contributed by veterans in the area which can be utilized by students and researchers alike.

A broad perspective, encompassing lucid descriptions of crop management towards higher yield and sustenance has been provided to the readers in this book. Agricultural production has a relation with physical problems which may not always be overcome by technology. However, one can see better groomed farms flourishing and consistently making greater profits as compared to similarly structured, neighboring farms under similar conditions. For every abiotic condition, it is well-known that there is a difference between the potential and observed yields, which is generally high and often could be cut down through better management techniques. This book discusses agricultural problems encountered in different regions of the world which were resolved using innovative methods, presenting new approaches for well-known techniques and new tools for old problems.

Each chapter is a sole-standing publication that reflects each author´s interpretation. Thus, the book displays a multi-facetted picture of our current understanding of application, resources and aspects of the field. I would like to thank the contributors of this book and my family for their endless support.

**Editor**

# Part 1

## Controlling Genome, Plant and Soil

# 1

# Population Management for Yield Improvement in Upland Rice Ecologies for West Africa Region

Andrew A. Efisue
*Department of Crop and Soil Science, Faculty of Agriculture*
*University of Port Harcourt*
*Choba, Port Harcourt*
*Nigeria*

## 1. Introduction

Rice has become a commodity of great importance in West and Central Africa (WCA), due to rural-urban migration, population growth and rapid urbanisation. These factors could be responsible for sudden change in the dietary preferences of consumers. Rice consumption rate was estimated to be growing by 5.6 % per annum, more than double the rate of population growth in WCA (WARDA, 1997). Currently, rice ranks fourth as most important grain crop in Africa, behind maize, sorghum and millet and West Africa region dominates the continent production (DeVries and Toenniesen, 2001).

In West Africa, the rainfed upland is important rice ecology; it represents 57% of total rice area and 40% of regional production with a potential yield of 2.5 to 4.5 tonnes/ha. However, in the farmers' field, yield realised is often very poor, not more than 1 tonne/ ha (Efisue et. al. 2008).

Apart from biotic and abiotic stresses attributed to farmers' low yield; there are inherent low-yield and poor agronomic characters in most of the rice cultivars currently in the farmers' fields. Although, high yielding Asian rice (*O. sativa)* is widely cultivated, it is very susceptible to biotic and abitioc stresses. The indigenous African rice (*O. glaberrima)*, unfortunately, has many undesirable agronomic characters such as grain shattering, long dormancy, and weak culms that lodge easily, which resultin low yield grain potential (Jones et al. 1997).

Farmers in West Africa still grow mixed proportion of these rice species (*O. glaberrima*, 15% and *O.sativa*, 20% ) of the total rice area (WARDA, 1993), while in some areas like Gao region of Mali a larger area is grown to *O. glaberrima* rice. The continuous growing of *O. glaberrima* might be due to some of its good agronomic characters, such as taste, aroma, and excellent vegetative growth which suppresses weeds (Dingkuhn et al., 1998). Jones et al. (1997) reported that *O. glaberrima* represents an invaluable reservoir of useful genes for biotic and abiotic stresses such as drought tolerance genes.

Thus, resource-poor farmers would appreciate developing an ideal variety that could perform well in any condition with higher yield potential. The major concern for rice breeders is to develop varieties with high yield potential, which is the result of several components that are determined at various stages in the growth of rice (Yoshida, 1981). Effective population management will enhance yield potential by selecting promising genotypes from the early generations of the rice crop, thus reduce the carry-over of undesirable materials into the next generation of the breeding cycle. The objectives of the study are to identify high yielding lines and desirable agronomic characters through population management practices for upland rice ecologies.

## 2. Materials and methods

### 2.1 Experimental site and weather conditions in Mali

The rice populations were established at the International Crops Research Institute for the Semi-Arid Tropics (ICRISAT) research station, Samanko, Mali in 2005 dry season. The soils were acidic and deficient in organic matter and total nitrogen. The soil texture was silty clay loam with very low cation exchange capacity. The soil was of low fertility. This resulted from inadequate levels of essential nutrients, especially Nitrogen (N) content of 0.058 %. The soil organic matter was low (0.480 %). Mali is located in sub-Sahelian vegetation belt in West Africa. It is a landlocked country located in the interior of West Africa between 12° W and 4° E longitude and 10 and 25° N latitude. The annual rainfall regime is monomodal, with distinct wet and dry seasons and air temperature very high during the early months of the year. The rainfall starts mainly in April and increases sharply in August, which is the peak period followed by sharp drop till October. The July and August receive about 60% of the annual rainfall, which shows the uneven distribution of rain in this region. The period between November and March (5 months) is virtually dry and no rain. This period also experiences the harmatan haze that blows from the Sahara desert to the Sahel region of West Africa. The mean monthly rainfall in the period under review was 86.12 mm. Mali has bimodal pattern in monthly air temperature and the air temperature increases from $34.8^0$ C in January to $43.5^0$ C in April, which is the hottest month. The second modal air temperature starts from August and increases gradually to November, and decreases thereafter till January.

### 2.2 Reference populations used for the experiment

The experiment comprised a total of 30 entries. They include 13 WBK populations derived from North Carolina design II mating scheme and were advanced to the F2 and F3 generations for seed increase by single plant selection. Eight lines (5 F6 and 3 F7) were selected at ICRISAT research station at Samako, Mali based on good agronomic characters from SIK 360 population received from Institut d' Economie Rurale (IER) at Sikasso Mali and 9 parental lines of these crosses (Table 1). The acrimony "WBK" is WARDA BAMAKO, while "SIK" is Sikasso.

## 3. Crop management and experimental design

The land was ploughed and disc harrowed and levelled before sowing the seeds. Dried seeds of rice were dibbled on shallow hole of 5 mm depth at the rate of three seeds per hole with a spacing of 20 cm within plants and 25 cm between rows and thinned to one plant per hole after 15 days of seedling emergency. The plot size was 1 m x 2 m (2 $m^2$) and a total of 44

| Population | Pedigree | Generation |
|---|---|---|
| WBK 35 | WAB 450-IBP-6-1-1/ WAB 365-B-1-H1-HB | F2 |
| WBK 39 | WAB 880-1-38-13-1-P1-HB / WAB 365-B-1-H1-HB | F2 |
| WBK 40 | WAB 880-1-38-13-1-P1-HB / WAB 375-B-9-H3-2 | F2 |
| WBK 41 | WAB 880-1-38-13-1-P1-HB / NERICA 2 | F2 |
| WBK 39 | WAB 880-1-38-13-1-P1-HB / WAB 365-B-1-H1-HB | F3 |
| WBK 40 | WAB 880-1-38-13-1-P1-HB / WAB 375-B-9-H3-2 | F3 |
| WBK 41 | WAB 880-1-38-13-1-P1-HB / NERICA 2 | F3 |
| WBK 42 | WAB 880-1-38-13-1-P1-HB / NERICA 3 | F3 |
| WBK 28 | WAB 450-IBP-103-HB / WAB 375-B-9-H3-2 | F3 |
| WBK 35 | WAB 450-IBP-6-1-1/ WAB 365-B-1-H1-HB | F3 |
| WBK 150 | WAB 880-1-38-13-1-P1-HB / NERICA 3// NERICA 2 | BC1F2 |
| WBK 106 | WAB 450-IBP-6-1-1/ NERICA 3// WAB 375-B-9-H3-2 | BC1F2 |
| WBK 64 | WAB 450-IBP-103-HB / WAB 375-B-9-H3-2// WAB 365-B-1-H1-HB | BC1F2 |
| SIK 360-2-1-3-15-B | TOG 5681/ BG 90-2 | F6 |
| SIK 360-1-9-1-3-B | TOG 5681/ BG 90-2 | F6 |
| SIK 360-1-B-3-1-11-B | TOG 5681/ BG 90-2 | F7 |
| SIK 360-1-B-2-1-2-B | TOG 5681/ BG 90-2 | F7 |
| SIK 360-2-1-4-13-B | TOG 5681/ BG 90-2 | F6 |
| SIK 360-1-13-1-2-B | TOG 5681/ BG 90-2 | F6 |
| SIK 360-1-11-1-1-B | TOG 5681/ BG 90-2 | F6 |
| SIK 360-1-B-1-1-4-B | TOG 5681/ BG 90-2 | F7 |
| WAB 880-1-38-13-1-P1-HB | Parental line | P |
| WAB 450-IBP-103-HB | Parental line | P |
| WAB 450-IBP-6-1-1 | Parental line | P |
| WAB 365-B-1-H1-HB | Parental line | P |
| WAB 375-B-9-H3-2 | Parental line | P |
| NERICA 2 | Parental line | P |
| NERICA 3 | Parental line | P |
| TOG 5681 | Parental line | P |
| BG 90-2 | Parental line | P |

Table 1. Rice populations, pedigree and generation of selection

plants stand per plot. Experimental design was an incomplete block of 5 x 6 rectangular lattice in two replications. Basal fertilizer was applied at the rate of 200 kg ha$^{-1}$ of 17-17-17, N-P-K and top-dressed in two splits at maximum tillering and flowering stage with urea (46 % N) at the rate of 100 kg ha$^{-1}$. Hand weeding was done prior to each fertilizer application and there was no preventive treatment especially against diseases and insect pests.

## 4. Measurements

Measurements were taken as at when due for all traits using Standard Evaluation System (SES) for the Rice Reference Manual (IRRI, 1996). Before the commencement of data taken, 10 plants were labelled at random from the middle rows of each plot and these plants were used for all data taken.

### 4.1 Tiller number and plant vigour

Tiller number was taken at maximum tillering stage of the plant and a total of ten samples of plants per plot were randomly taken for measurement, and plant vigour was taken at 45 days after germination.

### 4.2 Effective tiller number and plant height

Effective tiller was regarded as the tiller that bear panicle for harvest and counts were taken from the 10 labelled plants and tiller number per plant was taken at maximum tillering stage, which correspond to panicle initiation (PI) stage. Plant height was measured from soil surface to the tip of the tallest panicle on the 10 labelled plants.

### 4.3 Days to flowering

This was regarded as anthesis time when about 50% of the plants in each plot have flowered and panicles were fully exserted.

### 4.4 Grain yield

Rice plants were harvested when less than 5% of the grain husk turned tan colour, the 10 labelled plants were harvested individually and only full grains were considered for yield. The entire plants in the two middle rows were harvested and bulked for yield determination. The grains were air-dried and final yields were adjusted to 14 % moisture content and 1000-grain weight was calculated from the seed lots for each sample.

### 4.5 Plant biomass and harvest index

All the plants harvested from the two middle rows and the 10 individual plants were air dried (average daily air temperature 37 $^{0}$C) for three weeks for total dry matter determination and harvest index was (HI) calculated as grain yield per total dry matter, while panicle harvest index (PHI) was calculated as grain weight per panicle weight and grain to straw ratio was derived.

### 4.6 Panicle measurements

The following measurements were taken (1) number of full and empty grains per panicle was taken by removing the entire panicle grains and separated into full and empty grains, (2) the weight of the full and empty grains was taken. (3) Percentage sterility is referred to as the ratio of the empty grains to the total grains. (4) Panicle harvest index (PHI) was calculated as weight of full grain per panicle weight.

### 4.7 Statistical analysis

Analysis of variance (ANOVA) based on rectangular lattice design was performed for all measured traits using Statistical Analysis System (SAS, version 9.1, 2003) to test the significance of differences among genotypes.

# 5. Results

## 5.1 Effects of phenology on rice population management

Figure 1a is a cross between two interspecific lines, while Figure 1b and Figure 1c are crosses between interspecific and *O. sativa* lines (Table 1). Earliness is an important phenology in upland rice ecology, and selection for earliness among the rice populations could be important in rice production. It was observed that in the early generation of these crosses, larger proportion of the $F_2$ population skewed towards the early maturing parent (NERICA 2 and WAB 365 and WAB 6-1-1) while in later generation $F_3$, it skewed towards the late maturing parent (WAB 880). Due to the complex nature of these crosses there could be doubt of transgresive segregation in flowering. The $F_1$ progenies between *O.glaberrima* x *O. sativa* behave like *O. glaberrima* in most of the characters including flowering. This flowering behaviour between $F_2$ and $F_3$ in these crosses needs further investigation.

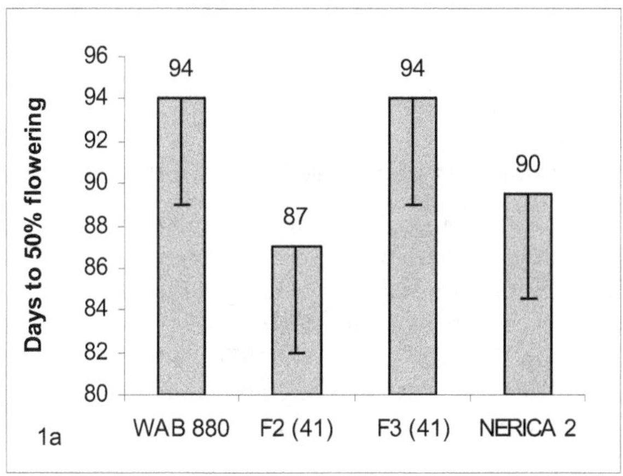


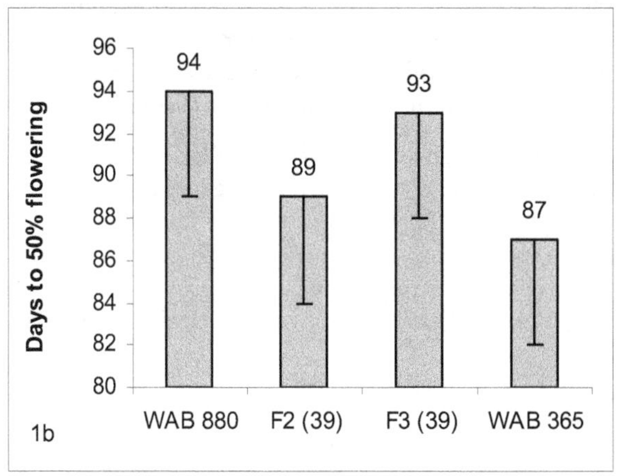

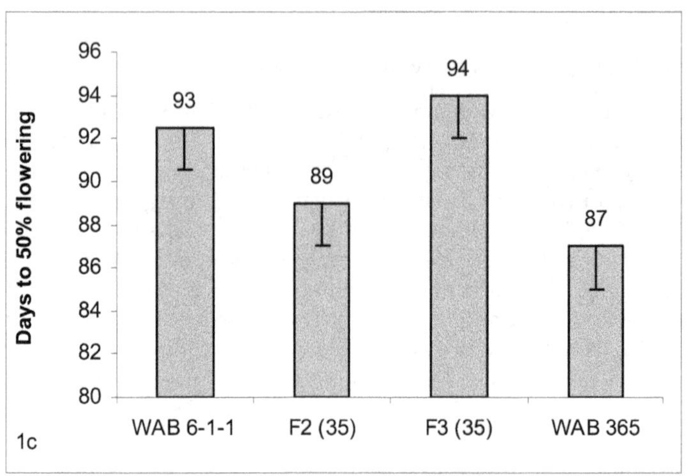


Fig. 1. Days to flowering of segregating populations (see Table 1 for populations code) .

## 6. Performance of populations and lines for some rice characteristics

The performance of rice populations and lines for some rice characteristics are presented in Table 2. Highly significant differences were observed among the rice populations for all the characters measured except the number of days to flowering ($P \leq 0.05$). Lines from SIK 360 population are photoperiod sensitive and their actual flowering dates could not be determined except for SIK 360-1-B-1-1-4-B and SIK 360-1-B-2-1-2-B now photoperiod-insensitive due to selections. The early maturing populations identified in this experiment are derived from either one of the early maturing parents such as WAB 365-B-1-H1-HB or NERICA 2 (Table 2).

The mean shoot dry weight of the F6 lines from SIK 360 populations was 626.88 g and F7 was 321.3 g, with a percentage decrease of 48.75% from one generation of the breeding cycle. Loss in shoot biomass was observed between $F_2$ and $F_3$ populations in all the crosses. This therefore raised concern on population management for rice production based on shoot biomass. The parental lines, TOG 5681 and BG 90-2 were higher in shoot dry weight as compared to the other parents including their progenies (Table 2).

Progenies from SIK 360 population were generally taller, with higher tillering ability and more vigorous than all the entries. These characters skewed towards the *O. glaberrima* parent (TOG 5681) except tiller number that skewed towards BG 90-2. Thus, this could indicate transgressive segregation for these characters and this phenomenon was not observed in the other crosses. Plant height for other populations ranged from 1.0 m to 1.3 m and with tiller numbers per plant ranging from 3 to 7, which could be ideal for the upland rice ecology.

## 7. Yield components

All yield components observed showed a significant variations among the rice populations (Table 3). High percent sterility was recorded in SIK 360 population and SIK 360-1-13-1-2-B

| Populations/lines | Generation | Days to Flowering | Shoot dry weight (g) | Plant height (cm) | Tiller number per plant | Vigour |
|---|---|---|---|---|---|---|
| WAB 365-B-1-H1-HB | Parent | 87 | 164.9 | 103.8 | 5.5 | 2.0 |
| WBK 41 | F2 | 87 | 218.1 | 108.2 | 5.5 | 3.0 |
| WBK 39 | F2 | 89 | 257.3 | 121.8 | 7.0 | 3.0 |
| WBK 40 | F3 | 89 | 233.8 | 116.3 | 5.0 | 6.0 |
| WBK 35 | F2 | 89 | 203.3 | 111.1 | 5.5 | 2.0 |
| NERICA 2 | Parent | 90 | 224.5 | 102.0 | 5.5 | 6.0 |
| WBK 64 | BC1F2 | 91 | 197.6 | 109.8 | 5.5 | 5.0 |
| WBK 106 | BC1F2 | 91 | 208.4 | 117.1 | 5.5 | 5.0 |
| WBK 150 | BC1F2 | 91 | 152.1 | 112.3 | 3.5 | 6.0 |
| WBK 39 | F3 | 93 | 164.6 | 122.7 | 5.5 | 6.0 |
| WAB 450-IBP-6-1-1 | Parent | 93 | 244.1 | 116.9 | 5.5 | 5.0 |
| WBK 35 | F3 | 94 | 166.7 | 118.8 | 5.5 | 4.0 |
| WAB 880-1-38-13-1-P1-HB | Parent | 94 | 243.8 | 105.2 | 6.5 | 5.0 |
| WBK 41 | F3 | 94 | 174.3 | 110.9 | 6.0 | 6.0 |
| WAB 450-IBP-103-HB | Parent | 95 | 151.6 | 132.3 | 4.0 | 7.0 |
| NERICA 3 (check) | Line | 96 | 196.4 | 116.7 | 5.0 | 4.0 |
| WBK 40 | F2 | 96 | 225.7 | 126.9 | 6.0 | 6.0 |
| WBK 28 | F3 | 96 | 171.7 | 121.4 | 4.5 | 5.0 |
| WBK 42 | F3 | 100 | 197.8 | 118.7 | 4.0 | 5.0 |
| SIK 360-1-B-1-1-4-B | F7 | 112 | 333.1 | 115.5 | 15.0 | 1.0 |
| SIK 360-1-B-2-1-2-B | F7 | 112 | 182.9 | 142.9 | 7.0 | 6.0 |
| BG-90-2 | Parent | 115 | 324.6 | 94.5 | 14.5 | 3.0 |
| SIK 360-1-11-1-1-B | F6 | PS | 622.0 | 132.0 | 20.5 | 2.0 |
| SIK 360-1-9-1-3-B | F6 | PS | 612.2 | 149.6 | 14.5 | 1.0 |
| SIK 360-1-B-3-1-11-B | F7 | PS | 448.0 | 159.7 | 9.5 | 2.0 |
| SIK 360-2-1-4-13-B | F6 | PS | 479.9 | 123.5 | 14.5 | 1.0 |
| SIK 360-2-1-3-15-B | F6 | PS | 629.8 | 137.2 | 11.5 | 1.0 |
| SIK 360-1-13-1-2-B | F6 | PS | 790.5 | 159.0 | 12.5 | 2.0 |
| TOG 5681 | Parent | PS | 472.7 | 140.1 | 21.0 | 1.0 |
| LSD (0.05) | | 9.3 | 158.66 | 19.21 | 7.19 | 2.47 |
| Probability | | * | *** | *** | *** | *** |

* significant at 0.05 probability level
** significant at 0.01 probability level
*** significant at 0.001 probability level
ns = no significance
PS = Photoperiod sensitive

Table 2. Mean performance of rice populations for some rice characters

had the highest value of 91.51% and WBK 35 with lowest value of 7.41%. All rice populations showed a significant ($P \leq 0.05$) response of effective panicles per square meter, which is higher in SIK 360 lines as compared to the WBK populations. Effective panicles are panicles that bear harvested rice grain. The WBK populations were significantly higher in 1000 grains weight as compared to SIK lines and none of the SIK lines were above the check. The three top best in 1000 grains weight are WAB 880-1-38-13-1-P1-HB, WBK 40 (F3) and WBK 40 (F2) and least is TOG 5681. A significant ($P \leq 0.001$) variation was observed for harvest index (HI) in all the rice populations. The results of harvest index followed similar pattern as with 1000-grain weight among the rice populations. Two parental lines that had the highest harvest index were WAB 365-B-1-H1-HB and NERICA 3 and as well as their progenies, while the SIK 360 lines were low (Table 3). Grain yield significantly varied among the populations and WBK 39 is the highest yielder of 4.35 t ha$^{-1}$. Six of the entries yielded higher than NERICE 3 (check) and TOG 5681 exhibited the lowest yield.

Higher percentage of sterility was observed in the progenies between crosses of interspecifics than with the cross between intserpecific x *O. sativa* (Figure 2a and 2b). The F3 progenies of the rice populations were observed to be more sterile than their parents and F2 progenies. The F3 progenies were 13.4% and 27.3% more sterile than most sterile parents in the crosses of intersepcific x *O. sativa* (Figure 2a) and interspecific x interspecific (Figure 2b), respectively. This might be due to cytoplsmic differences in each of the crosses. While in the cross between TOG 5681 x BG 90-2, the fertility of progenies increase gradually in each generation due to selection biased toward full grains (Figure 2c). The fertility difference between BG 90-2 and the F7 hybrid is 38.7%, thus, showed that interspecific hybrids require longer generations of selection to restore full fertility. This information has great implication for the development of high yielding interspecific hybrids.

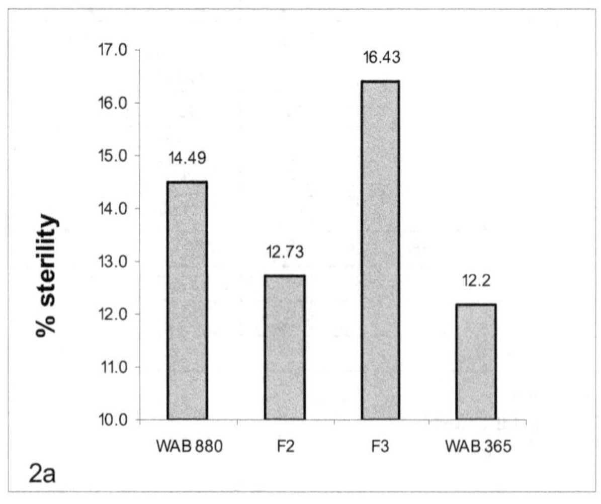

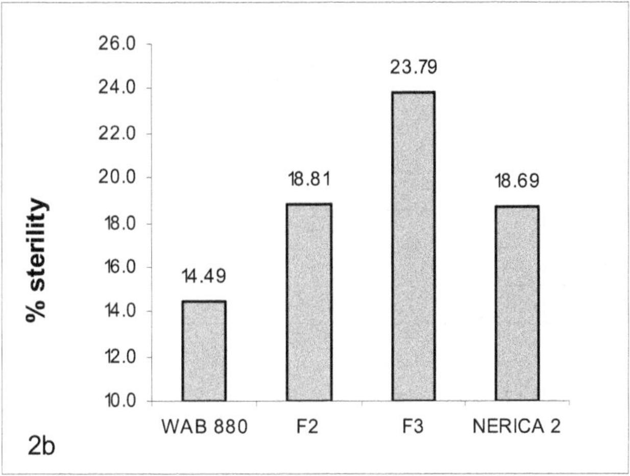


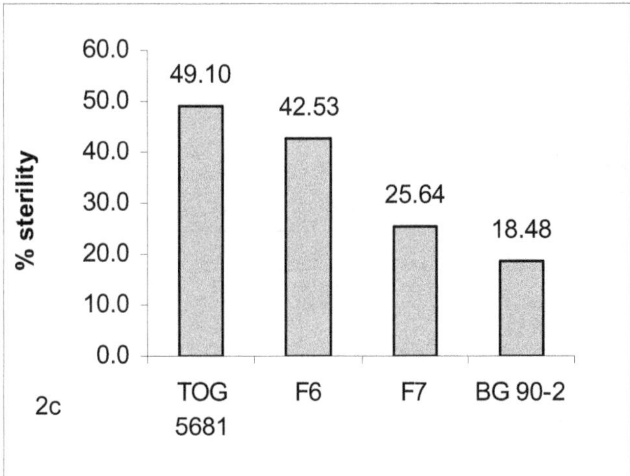


Fig. 2. The effect of sterility in each breeding cycle

## 8. Relationships among rice traits

Highly significant relationships were observed among all traits except with number of panicles per square meter (Table 4). High grain yield, harvest index and grain to straw ratio were found to be significantly ($P \leq 0.001$) associated with low spikelet sterility. Grain yield had no significant relationship with plant height and tiller numbers at panicle initiation stage. The non-significant relationship observed between tiller number and grain yield could be due to the fact that some tillers died before reaching maturity and also that some tillers were not effective in producing panicles. Percentage sterility was strongly and negatively correlated with yield components examined (Table 4).

| Populations/lines | Generation | % Sterility | Number of effective panicles (m2) | 1000 grain weight/g | Harvest index | Yield ( t/ha) |
|---|---|---|---|---|---|---|
| WBK 39 | F2 | 12.73 | 179 | 32.19 | 0.47 | 4.35 |
| BG 90-2 | Parent | 18.48 | 198 | 29.19 | 0.41 | 4.18 |
| SIK 360-1-B-1-1-4-B | F7 | 20.24 | 271 | 25.14 | 0.38 | 3.85 |
| SIK 360-1-11-1-1-B | F6 | 16.54 | 188 | 25.38 | 0.28 | 3.74 |
| SIK 360-1-B-2-1-2-B | F7 | 35.36 | 193 | 28.19 | 0.41 | 3.71 |
| WBK 40 | F3 | 15.68 | 179 | 34.54 | 0.48 | 3.69 |
| NERICA 3 (check) | Parent | 9.19 | 139 | 28.60 | 0.50 | 3.63 |
| WBK 39 | F3 | 16.43 | 164 | 31.30 | 0.49 | 3.56 |
| WBK 40 | F2 | 19.72 | 162 | 34.54 | 0.47 | 3.51 |
| SIK 360-1-9-1-3-B | F6 | 19.53 | 172 | 28.57 | 0.26 | 3.46 |
| WBK 35 | F3 | 19.59 | 159 | 28.77 | 0.47 | 3.34 |
| WAB 880-1-38-13-1-P1-HB | Parent | 14.49 | 125 | 34.86 | 0.48 | 3.30 |
| WBK 35 | F2 | 7.41 | 140 | 28.63 | 0.47 | 3.23 |
| WBK 64 | BC1F2 | 18.52 | 137 | 28.77 | 0.47 | 3.23 |
| WBK 42 | F3 | 14.68 | 146 | 33.40 | 0.46 | 3.22 |
| WBK 28 | F3 | 17.95 | 192 | 33.03 | 0.45 | 3.20 |
| WAB 365-B-1-H1-HB | Parent | 12.20 | 141 | 26.24 | 0.50 | 3.17 |
| WBK 106 | BC1F2 | 14.27 | 134 | 29.08 | 0.48 | 3.07 |
| SIK 360-1-B-3-1-11-B | F7 | 21.31 | 142 | 22.56 | 0.25 | 3.07 |
| NERICA 2 | Line | 18.69 | 171 | 25.83 | 0.45 | 3.06 |
| SIK 360-2-1-4-13-B | F6 | 28.27 | 168 | 23.42 | 0.28 | 3.01 |
| WAB 450-IBP-6-1-1 | Parent | 21.07 | 147 | 27.46 | 0.43 | 3.01 |
| SIK 360-2-1-3-15-B | F6 | 42.88 | 245 | 25.34 | 0.20 | 2.84 |
| WAB 450-IBP-103-HB | Parent | 14.09 | 103 | 32.96 | 0.44 | 2.79 |
| WBK 41 | F2 | 18.81 | 185 | 30.14 | 0.41 | 2.67 |
| WBK 150 | BC1F2 | 21.62 | 153 | 31.05 | 0.45 | 2.66 |
| WBK 41 | F3 | 23.79 | 196 | 27.97 | 0.43 | 2.65 |
| SIK 360-1-13-1-2-B | F6 | 91.51 | 159 | 20.88 | 0.09 | 1.39 |
| TOG 5681 | Parent | 49.10 | 139 | 19.18 | 0.05 | 0.50 |
| LSD (0.05) | | 19.400 | 76.903 | 2.550 | 0.068 | 1.315 |
| Probability | | *** | * | *** | *** | ** |

* significant at 0.05 probability level
** significant at 0.01 probability level
*** significant at 0.001 probability level

Table 3. Performance of rice genotypes for some yield components

| Traits | Yield | 1000gwt | HI | %Sterility | G-S ratio | No.Panicle/ m2 | PH |
|---|---|---|---|---|---|---|---|
| 1000gwt | 0.375** | | | | | | |
| HI | 0.576*** | 0.723*** | | | | | |
| %Sterility | -0.620*** | -0.495*** | -0.720*** | | | | |
| G-S ratio | 0.519*** | 0.694*** | 0.982*** | -0.675*** | | | |
| No.Panicle/m2 | 0.352** | -0.247 ns | -0.160 ns | 0.057 ns | -0.229 ns | | |
| PH | -0.12 ns | -0.345** | -0.631*** | 0.457*** | -0.612*** | 0.092 ns | |
| TN | -0.095 ns | -0.529*** | -0.676*** | 0.282* | -0.674*** | 0.250 ns | 0.395*** |

* *significant at 0.05 probability level*
** *significant at 0.01 probability level*
*** *significant at 0.001 probability level*
*ns: not significant*

Note: Yld= grain yield, 1000 grain weight, HI= harvest index, %Sterility, Grain to straw ratio, Pani/m2=Number of panicles /m2, PH= plant height and TN=tiller number at PI stage.

Table 4. Correlation coefficients among traits studied in 30 rice genotypes during 2005 wet season.

## 9. Performance of single plant progenies

Ten single plant progenies (SPP) were sampled within each plot and analysed (Table 5). Lines from SIK 360 population were higher in shoot dry weight than other populations and SIK 360-1-13-1-2-B had the highest value of 62.88 g. However, SIK 360 lines were lower in grain to straw ratio, harvest index (HI) and panicle harvest index (PHI) than other rice populations. There was significant variation among the populations in response to grain yield. There were 15 entries that yielded above grand mean yield (18.68 g) and six entries that yielded above the check. There was significant relationship between SPP yield and HI (r = 0.51***), which could assist in determining the performance of a population at early generations. Plot yield was significantly (P≤ 0.001) associated with SPP grain yield and harvest index (Figure 3a and 3b).

## 10. Discussion

### 10.1 Phenology in rice population management

Rice phenology plays an important role in grain yield determination in rice production. Rice varieties of appropriate phenology such as early maturing could be used to avoid adverse drought stress most especially, late season drought ( Fukai, 1999 and Fukai and Cooper, 1995). Early flowering populations identified in this experiment were WBK 35, WBK 39,

| Populations | Generation | Shoot dry weight /g | Grain: Shoot ratio | Panicle harvest index | Harvest index | Yield /g |
|---|---|---|---|---|---|---|
| SIK 360-1-B-2-1-2-B | F7 | 37.84 | 0.80 | 0.92 | 0.44 | 30.88 |
| WBK 39 | F2 | 26.81 | 0.97 | 0.95 | 0.48 | 25.24 |
| SIK 360-2-1-3-15-B | F6 | 57.89 | 0.53 | 0.90 | 0.30 | 24.41 |
| WBK 35 | F3 | 23.08 | 1.05 | 0.93 | 0.48 | 23.94 |
| SIK 360-1-B-1-1-4-B | F7 | 34.91 | 0.68 | 0.92 | 0.40 | 23.59 |
| WBK 40 | F3 | 22.00 | 1.09 | 0.92 | 0.52 | 22.70 |
| NERICA 3 (check) | Parent | 19.87 | 1.15 | 0.95 | 0.53 | 22.67 |
| WBK 28 | F3 | 24.85 | 0.93 | 0.92 | 0.48 | 22.61 |
| WBK 39 | F3 | 23.82 | 0.96 | 0.93 | 0.49 | 22.18 |
| BG 90-2 | Parent | 32.55 | 0.69 | 0.93 | 0.40 | 21.81 |
| WBK 35 | F2 | 19.00 | 1.13 | 0.95 | 0.53 | 21.03 |
| WBK 40 | F2 | 20.32 | 1.03 | 0.94 | 0.49 | 20.12 |
| WBK 42 | F3 | 20.22 | 0.97 | 0.96 | 0.48 | 19.86 |
| WAB 450-IBP-103-HB | Parent | 22.96 | 0.82 | 0.94 | 0.44 | 19.80 |
| WAB 365-B-1-H1-HB | Parent | 18.20 | 1.10 | 0.93 | 0.50 | 18.53 |
| WBK 41 | F3 | 20.65 | 0.88 | 0.92 | 0.46 | 17.90 |
| WBK 106 | BC1F2 | 15.93 | 1.12 | 0.93 | 0.52 | 17.73 |
| SIK 360-1-B-3-1-11-B | F7 | 51.42 | 0.36 | 0.88 | 0.26 | 17.72 |
| WBK 64 | BC1F2 | 18.96 | 0.96 | 0.93 | 0.48 | 17.52 |
| WAB 450-IBP-6-1-1 | Parent | 18.08 | 0.98 | 0.94 | 0.49 | 17.45 |
| SIK 360-1-11-1-1-B | F6 | 40.89 | 0.43 | 0.94 | 0.30 | 17.21 |
| WBK 150 | BC1F2 | 20.60 | 0.86 | 0.93 | 0.44 | 17.12 |
| NERICA 2 | Parent | 19.50 | 0.81 | 0.91 | 0.45 | 16.22 |
| WBK 41 | F2 | 19.99 | 0.83 | 0.92 | 0.44 | 15.38 |
| WAB 880-1-38-13-1-P1-HB | Parent | 15.24 | 1.13 | 0.91 | 0.46 | 14.56 |
| SIK 360-1-9-1-3-B | F6 | 46.13 | 0.35 | 0.93 | 0.25 | 14.41 |
| SIK 360-2-1-4-13-B | F6 | 35.52 | 0.45 | 0.92 | 0.29 | 14.14 |
| SIK 360-1-13-1-2-B | F6 | 62.88 | 0.11 | 0.83 | 0.08 | 5.20 |
| TOG 5681 | Parent | 59.06 | 0.04 | 0.64 | 0.04 | 2.17 |
| LSD (0.05) | | 16.844 | 0.2927 | 0.0373 | 0.0901 | 9.9103 |
| Probability | | *** | *** | *** | *** | ** |

* significant at 0.05 probability level
** significant at 0.01 probability level
*** significant at 0.001 probability level
ns : no significant

Table 5. The performance of single plant progenies in the rice populations.

WBK 40 and WBK 41, which could be suitable for late season drought and may also promote second cropping in some environments. These short duration populations performed favourably in relation to yield as compared to late maturing populations. Their

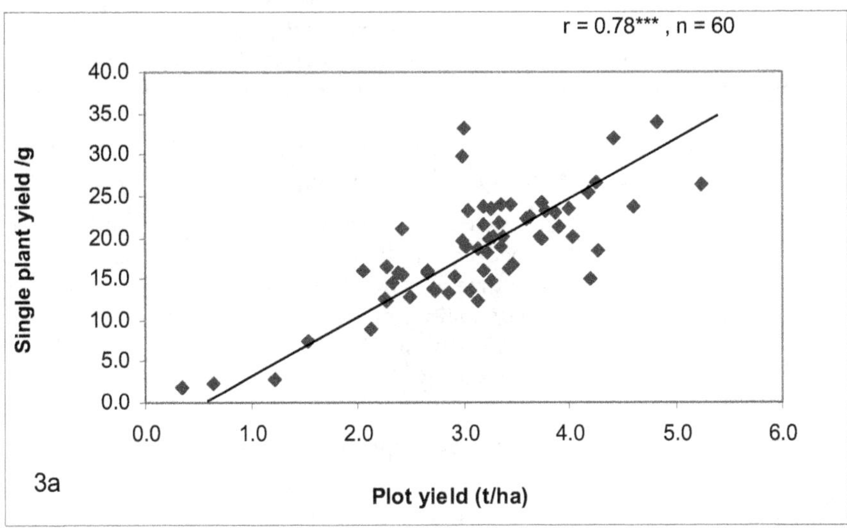


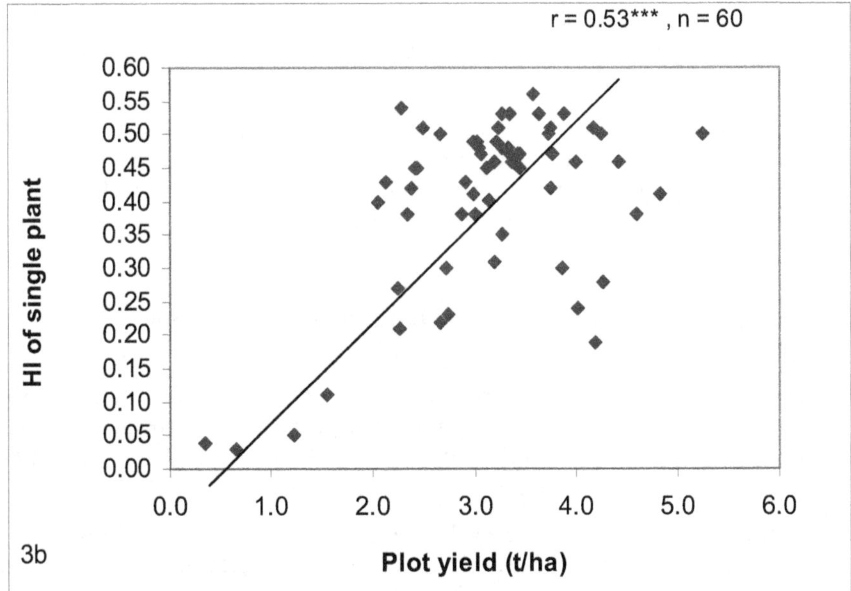


Fig. 3. Relationship between plot yield vs SPP grain yield (3a) and HI (3b) from all populations

favourable yield performance might be attributed to their high harvest index as they are often more efficient in nutrient use than varieties with lower harvest index ( Inthapanya et al. 2000). The flowering pattern observed in some of the crosses at each generation might be a good signal in selections to suit differential seasonal planting in some rice ecologies, as farmers do engage in one or more business ventures. This flowering pattern (Figure 1a, 1b and 1c) could be attributed to either effect of epistasis that masked late maturing genes or dominance of early maturing genes in the early generations. As breeding cycle advances, genetic manipulations such as crossing over might occur that unmasked some hidden genes resulting in higher level of segregants (Falconer and Mackay, 1996). This shows that high selection pressure may not be applicable with interspecific hybrids for selecting early maturing materials at the early generations of the breeding cycle.

Most traditional rain-fed lowland rice cultivars are sensitive to photoperiod and later maturing (Mackill et al. 1996). It is believed that they produced higher grain yield than the early maturing rice cultivars as more time is allowed for the plants to utilize more of the available resources and better recovery ability from the early drought (Fukai, 1999). Thus, SIK 360 lines, WBK 39, WBK 40, WAB 450-IBP-6-1-1 and WAB 880-1-38-13-1-P1-HB are identified as potential materials for this ecology.

## 11. Major agronomic characters in rice population management

The high expressivity of earliness observed in some of the rice populations from early maturing parents ( NERICA 2 and WAB 365-B-1-H1-HB) indicates importance of early maturing donors in breeding for early maturity in rice. In West Africa, some communities use rice straws as supplements for animal feeds. Therefore, the loss of shoot biomass from successive generations in breeding cycles implicate rice breeders who will breed for rice straw for these communities in West Africa.

Populations such as SIK 360 lines, WBK 39, WBK 40, WAB 450-IBP-6-1-1 and WAB 880-1-38-13-1-P1-HB could be potential materials for weed competitiveness as they possess high seedling vigour, tillering ability and shoot dry weight characters, which significantly correlate with weed competitiveness in rainfed rice ecologies (Fofana et al. 1995; Dingkuhn et al. 1998).

## 12. Yield components and associated characters in rice population management

Grain weight is a veritable parameter in rice, in comparison with other cereals (Yoshida, 1981) and 1000 grain- weight is significantly associated with yield components that could be exploited for higher yield. Progenies from WBK populations were higher than the SIK 360 lines in the yield components examined.

Spikelet sterility is significantly negatively correlated with grain yield, as significant variations were observed among the rice populations. Thus, selection of low percentage spikelet sterility in each generation as an indirect method for grain yield selection, which may save time and could be cost effective, this is consistent with the work of Garrity and O'Toole (1994), who found that fertility is related to grain yield in plants that are exposed to drought during flowering. In a related experiment, Lafitte et al., (2006) reported that yield was closely correlated with spikelet fertility (estimated from panicle harvest index (PHI). Yield components in single plant progeny were significantly associated with plot yields, and this could be a rapid way of assessing the performance of rice population in the early generation of the breeding cycle thereby reducing the numbers of material to carryover into the next generation. Lafitte et al., (2006) suggested that highly heritable characters showing high correlations with single plant yield might be used as an indirect selection for yield in early generations such as harvest index for single plant in some of the rice populations.

## 13. Conclusion

The early maturing rice populations (WBK 35, WBK 39, WBK 40 and WBK 41) identified in this experiment could be deployed in early drought prone environments while the late maturing populations (SIK 360, WBK 42 and BG 90-2) could be suitable for intermittent or terminal drought prone environments. The deployment of these populations to these water stressed environments will increase rice production in the region. The flowering patterns observed, which was attributed to cytoplasmic differences of the parental lines indicates that interspecific hybrids require longer generations of selection to restore full fertility. This information has great implication in the development of high yielding interspecific hybrids. Rice population management is an effective way in fast-tracking development of rice to be deployed to farmers in the region.

## 14. References

Dingkuhn, M., M.P.Jones, D.E. Johnson & A. Sow, 1998. Growth and yield potential of Oryza sativa and Oryza glaberrima upland rice cultivars and their interspecific progenies. *Field Crop Res.* 57: 57-69.

DeVries, J. and Toenniessen, J. 2001. Securing the harvest. *Biotechnology, breeding and seed systems for African crops.* CABI Publishing, Wallingford, Oxon OX10 8DE, UK, 208pp.

Efisue, A., Tongoona, P., Derera, J., Langyintuo, A., Laing, M.and Ubi, B. (2008). Framer,s perceptions on rice varieties in Sikasso region of Mali and their implications for rice breeding. *J. Agronomy & Crop Science* 194: 393-400.

Falconer, D.S. and Mackay, T.F.C. 1996. Introduction to quantitative genetics, 4th edition. Pearson Prentice Hall. ISBN 0-582-24302-5, pp 464

Fofana, B., Koupeur, T., Jones, M.P. and Johnson, D.E. 1995. The development of rice varieties competitive with weeds. *Proceedings of the Brighton Crop Protection Conference,* 187-192.

Fukai, S. 1999. Phenology in rainfed lowland rice. *Field Crops Res.* 64: 51-60

Fukai, S. and Coop, M. (1995). Development of drought-resistance cultivars using physio-morphological traits in rice . *Field Crops Res.* 40: 67-86.

Garrity, D.P. and O'Toole, J.C. 1994. Screening rice for drought resistance at the reproductive phase. *Field Crops Res.* 39: 99-110

Inthapanya, P., Sipaseuth, , Sihavong, P., Sihathep, V., Chanphengsay, M., Fukai, S. and Basnayake, J. 2000.Genotype differences in nutrient uptake and utilization for grain yield production of rainfed lowland rice under fertilization and non-fertilized conditions. *Field Crops Res.* 65: 57-68

IRRI., 1996. Standard Evaluation System for Rice. International Rice Research Institute, Los Baanos, Philippines.

IRRI, 1996. Rainfed lowland rice improvement. D.J Mackill, W.R. Coffman and D.P. Garrity (eds.). pp 242

Jones, M.P., Dingkuhn, M., Aluko, G.K. and Semon, M. 1997a. Using backcrossing and double haploid breeding to generate weed competitive rice from *O.sativa* L. x *O.glaberrima* Steud. genepools. In: *Interspecific hybridization: Progress and Prospects,* Edited by M.P. Jones , M. Dingkuhn, D.E. Johnson and S.O. Fagade. *Proceedings of the workshop: Africa / Asia Joint Research on Interspecific Hybridization between the Africa and Asian Rice Species (O. glaberrima and O. Sativa).* WARDA, Mbe, Bouake, Cote d'Ivoire, Dec. 16-18, 1996. pp 61-79

Jones, M.P., Dingkuhn, M., Aluko, G.K. and Semon, M. 1997b Interspecific O.sativa L. x O. glaberrima Steud. Progenies in upland rice improvement. *Euphytica,* 92: 237-246.

Lafitte, H.R., Li,Z.K., Vijayakumar, C.H.M., Gao, Y.M., Shi, Y., Xu, J.L., Fu,, B.Y., Yu, S.B., Ali, A.J., Domingo, J., Maghirang, R., Torres, R. and Mackill, D. (2006). Improvement of rice drought tolerance through breeding: Evaluation of donors and selection in drought nurseries. *Field Crops Res.* 97: 77-86.

Mackill, D.J. Coffman, W.R. and Garrity, D.P. 1996. Rainfed lowland rice improvement. IRRI, Manila, Philippines, pp242.

SAS Institute Inc. 2003. SAS / STAT user's guide, version 9.1. SAS Institute Inc., Cary, NC., USA.

WARDA 1997. *Medium Term Plan 1998-2000: For presentation to the Technical Advisory Committee (TAC), Consultative Group on International Agricultural Research, Rome, Italy.* WARDA, Bouake, Cote d'Ivoire.

WARDA 1997. *Medium Term Plan 1998-2000: For presentation to the Technical Advisory Committee (TAC), Consultative Group on International Agricultural Research, Rome, Italy.* WARDA, Bouake, Cote d'Ivoire.

WARDA, 1993. Annual Report, Bouake, Cote d'Ivoire

Yoshida, S. 1981. Fundamentals of rice crop science. The International Rice Research Institute (IRRI). Los Banos, Laguna, Philippines. pp 269

2

# Ripening and the Use of Ripeners for Better Sugarcane Management

Marcelo de Almeida Silva[1] and Marina Maitto Caputo[2]
[1]*São Paulo State University / UNESP – College of Agricultural Sciences*
[2]*University of São Paulo / USP –"Luiz de Queiroz" College of Agriculture*
*Brazil*

## 1. Introduction

The sugarcane crop is important because of its multiple uses: it is used around the world "in natura" for forage for animal feed or as raw material for the manufacture of brown sugar, molasses, alcoholic beverages, sugar and ethanol. Processing its waste also has great economic importance, especially for vinasse, which is transformed into organic fertilizer, and for bagasse, which is transformed into fuel. Sugarcane is grown mainly in tropical and subtropical climates across a wide region, between the 35° north and south latitudes from the equator. The ideal climate has two distinct seasons: the first is hot and humid to facilitate budding, tillering and plant growth and the second is cold and dry to promote maturation and the consequent accumulation of sucrose in stems.

Distinct from the plants grown at the beginning of commercial exploitation, there are now genetically improved varieties of sugarcane that have low levels of fiber and high concentrations of sugar, which are both responsible for high productivity. However, despite the diversity of genetic material available, the sugarcane industry still faces problems with the precocity of raw material, and it is not capable of meeting industry demand at the beginning of the season as is it during the middle of the season to produce the same amount of material required for processing. Another problem that has not been fully resolved for industrialization is flowering.

The process of flowering, an important aspect in the production of sugarcane, involves morpho-physiological alterations of the plant, which are considered highly undesirable features when accompanied by an intense pith process (the process of juice loss in the parenchyma cells of the stalk) and may modify the quality of the raw material from a technological point of view. Substantial losses in the productivity of stems and sugar are attributed to flowering during the season. The pith process in the stem begins with blooming, causing dehydration of the tissue and a consequent loss of weight in the stem; thus, it becomes very important to quantify the degree of pith processing and the possible modifications in the quality of the raw material to properly design the area to be planted for each variety and to determine the time periods most favorable for their industrialization. However, depending on the variety and environmental conditions to which it is subjected, the intensity of the process is variable, as is the intensity of the problems arising from these phenomena.

A gradual drop in temperature and reduced rainfall are crucial for the natural ripening of sugarcane. However, in many years during the ripening period, these conditions do not occur simultaneously or completely. In such situations, the use of ripeners and flowering inhibitors in the sugarcane crop is designed to increase productivity and to anticipate the harvest, thus allowing essential crop management in a modern production system.

Ripeners, defined as plant growth regulators, act by altering the morphology and physiology of the plant, which can lead to quantitative and qualitative changes in production. They may act by promoting a reduction in plant growth and enabling increases in sucrose content, early ripening and increased productivity, and they may also act on enzymes (invertases), which catalyze the accumulation of sucrose in the stalks. The use of ripeners in the sugarcane production system has provided greater flexibility in managing the harvest and is highly relevant for harvest planning. Additionally, ripeners promote the industrialization of a better quality of raw material. However, the feasibility of utilization depends on a number of factors, including climatic, technical and economic variables; feasibility especially depends on the additional responses that each variety may provide to this cultivation practice.

The number of studies involving flowering and ripeners has increased. Doses and products have been tested to achieve greater sugar productivity without causing damage to the plant or decreases in the agricultural productivity of the current year and subsequent ratoons.

The main chemicals used as ripeners include ethephon, glyphosate, trinexapac-ethyl, sulfometuron-methyl, fluazifop-p-butyl and others, such as maleic hydrazide, paraquat and imazapyr.

This chapter describes the alterations that occur in sugarcane during the process of sucrose accumulation and flowering and provides information on crop management options with the application of ripeners and chemical flowering inhibitors when climate conditions are not suitable for natural ripening.

## 2. Sugarcane ripening

The ripening of sugarcane is a characteristic of the plant that can be stimulated by environmental and management factors.

Fernandes (1982) defined sugarcane ripening as a physiological process that involves the synthesis of sugars in the leaves, translocation of the products formed and storage of sucrose in the stalk. According to Castro (1999), ripening is one of the most important aspects for sugarcane production; in the process of sucrose accumulation, which depends on energy, varietal characteristics are fundamental.

According to Deuber (1988), the ripening of sugarcane may be considered from three different points of view: botanical, physiological and economic. Botanically, sugarcane is ripe after the production of flowers and formation of seeds, which may give rise to new plants. Taking into account vegetative reproduction, which is the manner used by the productive sector, plants may be considered ripe much earlier in the cycle when the buds are ready and able to give rise to new plants. Physiologically, ripening is reached when the stems reach their potential sucrose storage, i.e., the point of maximum possible sucrose accumulation. In this cycle, the sugarcane reaches full botanical maturity before reaching physiological maturity. This means that although the seeds may already be falling in panicles, the accumulation of sucrose still continues for a period of generally one to two months. Economically, from the perspective of agronomic practice, sugarcane is considered mature or able to be industrialized from the moment it possesses the minimum sucrose content (pol), which is greater than 13% of the stalk weight.

For example, in the southeastern region of Brazil, which is considered to be the largest producer of sugarcane in the world, the culture's ripening process occurs naturally from the beginning of May and reaches its climax in October. The region's climatic conditions, which include a gradual drop in temperature and a decrease in precipitation until it is completely dry in the middle of the year, are decisive in the maturation process.

During ripening, sugarcane stores sucrose starting from the bottom and then in the top of the stalk. Therefore, at the beginning of the season, the lower third of the stalk has a higher sucrose content than the middle third, which has a higher content than the upper third. As ripening progresses, the sucrose content tends to equalize in the different parts of the stalk.

The increase in sucrose accumulation in the internodes of the already developed stalks is strongly influenced by environmental conditions that are unfavorable for plant growth and development. In this stage, 11 to 20 months after planting (depending on time of installation of the cane field, the period of ripening for the variety used and the country where it is being produced), there is full ripening of the sugarcane, which is observed when the harvest is properly monitored by specific technological analyses.

Varieties are classified as a function of whether they ripen in the early, mid or late stages; in other words, varieties are classed by when they reach the sucrose content suitable for industrial uses (i.e., at the beginning, middle or end of the season) without accounting for establishing the period of maximum sucrose content.

Worldwide, sugarcane generally has only one planting season; however, in Brazil, there are two distinct seasons for sugarcane planting. One-and-half-year planting, or sugarcane in 18 months, which is planted from January to early April, has a limited growth rate (zero or even negative), depending on the weather conditions from May to September. The phase of higher crop development takes place mainly from October to April, if there are good rainfall conditions, with peak growth from December to March. This is the planting period preferred by farmers. However, sugarcane planted in one-year cycle is usually planted from September to October, with its maximum development occurring between the months of November through March, which decreases after these months due to adverse weather conditions. The possibility of harvest begins in July, depending on the variety. This planting method is used for areas that are planned for the end of the harvest season.

The sugarcane ripening process has been studied in several countries. Given its relationship with agronomic experimentation, routine assessment of the stage of ripening and the cost of sugarcane based on sucrose content, it is of fundamental importance to understand the attributes in the juice and stalk, such as Brix, pol, juice purity, sugar cane fiber and reducing sugars, and the total recoverable sugars during plant development, (Caputo, 2006).

### 2.1 Mechanism of sucrose accumulation in sugarcane stalks

Sugarcane has the capacity to maximally use the available sunlight for photosynthesis. Each internode produces a new leaf in approximately ten days, and older leaves senesce, leaving a constant number of eight to nine leaves per stem. The majority of incident light is intercepted by the six most apical leaves (Alexander, 1973).

According to Legendre (1975), rapid fluctuations in temperature have little effect on the photosynthetic process, within the limits of 8-34 °C. Temperatures of 17-18 °C appear to be particularly favorable for the partition of photosynthates into the interior sugar reserves and the accumulation of high levels of sucrose.

However, according to Castro (2002), low temperatures lead to a rapid decline in photosynthetic efficiency, and high levels of sugar are not retained; the low temperatures affect the development of the stalk, sugar transport and storage, leading to accumulation of sucrose in the leaves. There is an interactive effect between sunlight, temperature and different varieties of sugarcane in response to the maturation process; however, with respect to rainfall, there is apparently no association with the referred process.

According to Moore (1995), the synthesis of sucrose that occurs in the cytosol and the synthesis of starch that occurs in the chloroplast are competitive processes that are established in sugarcane leaves. The metabolic pathways of sucrose and starch synthesis have several phases in common that involve certain enzymes, but these enzymes have isozymes that have different properties and are unique to an appropriate cellular compartment. The excess of triose phosphate can be used both for the synthesis of sucrose in the cytosol and for the synthesis of starch in the chloroplast, and the conditions that promote one inhibit the other.

According to Lingle (1999), sucrose, the end product of photosynthesis, passes through the phloem before being deposited in the vacuole and finally undergoes changes within storage cells. The inversion of sucrose, interconversion and phosphorylation of glucose and fructose, synthesis of sucrose phosphate and active accumulation through the tonoplast constitute the storage process. This process is characterized by the movement of sucrose against a concentration gradient. Through the inversion of fructose and glucose, sucrose can leave the vacuole once again.

Independent of the maturity of tissues, the mechanism of active sucrose accumulation appears to be the same, but there are differences between these tissues (mature and immature) with respect to the accumulation of sucrose due to the concentration of invertase and the need for growth. The immature storage tissues are characterized by cell expansion. The sucrose accumulated in these tissues is rapidly hydrolyzed by acid vacuolar invertase, and the hexoses produced move freely to the cytoplasm to be used in the growth process. As a part of the cycle, the hexoses may also be stored again.

In mature stem tissues, where growth processes are near completion, there is a decline in the concentration of acid vacuolar invertase, and neutral invertase then becomes predominant (the enzyme is apparently situated in the cytoplasm). This enzyme, together with cell wall acid invertase, governs the active accumulation of sucrose in the vacuole. Later, in more mature tissues that exhibit sucrose levels of approximately 15 to 20%, sucrose is stored in the intercellular spaces. Thus, sucrose plays an important role in the movement of sugars, depending on the physiological condition of plants, as conditioned by the environment; this may result in passive diffusion between the intercellular space and the vacuole by translocation to areas of intense sugar metabolism.

The synthesis of sucrose promotes the enrichment of stalk internodes, while the synthesis of starch reveals the immaturity of the sugarcane plant. Young plants and the apical region of the stalk are rich in starch, as are plants that have lost apical dominance due to the effect of the application of chemical products, improper management or unfavorable environmental factors.

In the elongation phase, assimilates are employed in the construction of the structure of internodes. Thus, the loss rate of assimilates is greater than the utilization rate because glucose, fructose and sucrose begin to be accumulated during this phase. Acid invertase and sucrose synthase reach their maximum activities in this process, demonstrating their association with this phase. Water content is negatively correlated with the activity of sucrose phosphate synthase, and the latter is positively correlated with sucrose content (Lingle, 1999).

## 2.2 The role of enzymes in ripening

Sucrose is accumulated in the stalks against a concentration gradient; the energy required for this process is provided by respiration. It has been established that increased sucrose content is accompanied by a continuous cycle of degradation and synthesis during the accumulation of sucrose in the reserve tissues (Vorster & Botha, 1999; Rohwer & Botha, 2001).

Many enzymes govern the primary metabolism of sucrose. Invertases have a key role in the partition of photosynthates between storage and growth, in which sucrose is broken down into glucose and fructose. These enzymes are classified by solubility, cellular location and optimum pH. The best-characterized isoforms are the acid invertases that occur in the apoplastic space, both free-form and bound to the cell wall. The soluble isoforms are predominantly present in the vacuole.

The activity of soluble acid invertase can be high or low, respectively, under favorable or unfavorable growth conditions (e.g., water stress, short photoperiod, low temperatures and application of ripeners).

Neutral or alkaline invertase occurs in the cytosol and still requires better characterization.

Sucrose-phosphate synthase (SPS) is considered to be the main regulatory enzyme in the pathway of sucrose synthesis. It synthesizes sucrose-6-phosphate, which is dephosphorylated by the action of the sucrose-phosphate phosphatase enzyme. Sucrose synthase (SuSy) can break down sucrose, generating UDP-glucose and fructose; it can also catalyze the reverse reaction, synthesis. In vivo, SuSy acts preferentially in the direction of the breakdown of sucrose.

According to Zhu et al. (1997), increasing the concentration of sucrose in individual sugarcane internodes is correlated with a decrease in soluble acid invertase activity during ripening.

The roles of soluble acid and neutral invertases (SAI and NI, respectively) and acid invertase are linked to cell wall mobilization, utilization and accumulation of sucrose in different varieties of sugarcane. The difference between sucrose concentrations in some genotypes is correlated with the difference between the activities of sucrose-phosphate synthase and soluble acid invertase enzymes. Ebrahim et al. (1998) and Lingle (1999) showed in their studies that both the activity of SPS and the difference between SPS and soluble acid invertase is correlated with the concentration of sucrose in developing internodes. Botha & Black (2000) demonstrated a positive correlation between SPS activity and the rate of sucrose accumulation and a highly significant correlation between SPS activity and sucrose content.

Environmental conditions can greatly influence the activity of invertase. Terauchi et al. (2000) reported that the activity of SAI decreased in cold conditions. The highest activity of SPS and the lowest activity of SAI probably resulted in an increase in the concentration of sucrose in the winter, at low temperatures, suggesting that the activity of SPS is one of the factors involved in the control of sucrose accumulation in sugarcane. Leite et al. (2009) observed an increase in SAI activity through the use of chemical ripeners under high precipitation, a condition favorable for sugarcane plant development. However, when subjected to decreases in temperature and precipitation, which are favorable conditions for natural ripening, the application of chemical ripeners resulted in elevated NI activity.

Lingle (1997) suggested that the activity of SAI was responsible for controlling growth in sugarcane plants. It was observed that the total concentration of sugar and sucrose increased while the activity of SAI decreased during the maturation of internodes, leading to the conclusion that this enzyme suppresses the accumulation of sugar.

The NI activity and sucrose content in mature internodes are closely related. Vacuolar SAI allows the accumulation and effective storage of sucrose when it is nearly absent (Suzuki, 1983). Rose and Botha (2000) also found a significant correlation between sucrose content and the level of NI.

Plant growth regulators promote distinct and significant alterations in the enzymatic activities of acid and neutral invertases (Leite et al., 2009, Siqueira, 2009).

### 2.3 Climatic factors that affect ripening

Sugarcane productivity is most influenced by climate (Barbieri, 1993; Keating et al., 1999).

Alexander (1973) mentioned that sugarcane plants slow their growth rate to accumulate more sugar under specific conditions involving both temperature and soil moisture content. The author also reported that the physiological ripening process limits the rate of plant growth without affecting the photosynthetic process, such that there is a greater balance of products that are photosynthesized and transformed into sugars for storage in plant tissues. Under these conditions, climate is defined as the major determinant of the restrictions imposed by the physical environment, which are represented by the interaction of climate factors that influence the soil and plant, such as the harvest season, the programmed number of plant cycles and the choice of crop varieties.

Air temperature plays a key role in sugarcane ripening: temperature is responsible for slowing the growth rate, leading to the accumulation of more sugar. Glover (1972) noted that low temperatures increased the sucrose content in stalks.

Sugarcane growth is also governed by genetic makeup and the environment. In general, the conditions of all seasons affect the development of sugarcane, and the success of the culture is intrinsically linked to environmental conditions favorable for its development.

Because ripening is the inverse of growth (Alexander, 1973), the method of negative-degree days is used to correlate ripening and temperature, corresponding to the area between the base temperature and the daily minimum temperature (Scarpari & Beauclair, 2004).

Since the 1940s, studies have shown that the variety of plant plays an important role in the ripening process because the different varieties have different ripening times, even when subjected to the same soil and climatic factors.

Sugar production (assimilation) is governed mainly by solar energy in the form of light and heat, while the use of sugars (dissimilation) depends largely on moisture and growth. The balance between production and use is reflected in the sucrose content of sugarcane. To ripen, the stalk must suffer growth retardation; low temperatures and moderate drought, among other factors, are effective agents for accelerating ripening. In tropical regions, moisture is essential for ripening, while in the subtropical regions, suitable minimum temperatures are essential.

## 2.4 Flowering of sugarcane

Sugarcane flowering (Fig. 1) has been regarded as harmful to the process of sucrose accumulation, as it is commonly accepted that the formation of flowers drains a considerable amount of sucrose. Substantial losses in sugarcane productivity and sucrose content at harvest are attributed to flowering.

Another aspect refers to the phenomenon of the pith process (Fig. 2) related to the flowering and ripening of sugarcane, which occurs in some varieties and is characterized by the drying of the interior of the stalk from the top. The pith process leads to a loss of moisture in the tissue, with a consequent reduction of the juice stock and an increase in the fiber content of the stalk. Thus, although the concentration of sucrose in the stalk is increased, it is difficult to extract and involves the loss of stalk weight and, consequently, a reduction in the final yield.

Fig. 1. Flowering of sugarcane.

Quantification of the degree of the pith process and the possible changes in the quality of the raw material may provide critical data to the design of the area to be planted for each variety and the best time periods for industrial use. The intensity of the flowering process and its consequences for the quality of raw materials vary with the variety and climate. Therefore, reducing the amount of juice is the main factor that is affected by flowering.

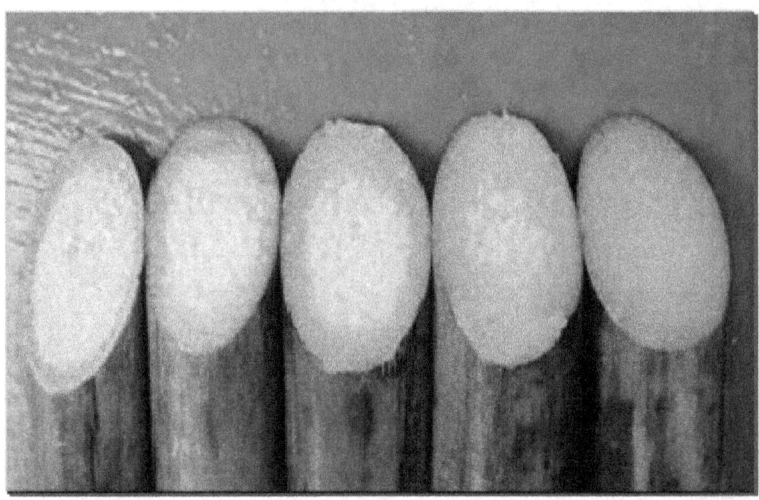

Fig. 2. Different grades of pith process in sugarcane. From the left to the right, intense pith to no pith.

The flowering process occurs simultaneously with the ripening process. Araldi et al. (2010) reported that the factors that influence flowering are the sensitivity of the variety to flowering, minimum age of the plant (varieties that are very sensitive to flowering may be induced at six months), photoperiod (optimum is approximately 12.5 hours) and light intensity, temperature (small variations in temperature can cause significant changes in flowering), humidity (humid and cloudy days favor flowering, which is less common in hot, dry regions), altitude (lower altitudes favor flowering), chemical products (various hormonal chemical products decrease flowering, which is of great practical interest) and fertilization (excess nitrogen can hinder or prevent flowering).

Sugarcane flowers during short days. In the Southern Hemisphere, the differentiation of the flower bud occurs from February to April, and the emergence of panicles occurs from April to July. In the Northern Hemisphere, these factors occur between July and August and from September to November, respectively.

Each variety has its particular length-of-day period, within which floral initiation can occur, as other factors, such as stage of development and nutritional aspects, are favorable. In some varieties, this time is broad, giving rise to abundant blooming for a considerable period; however, there are varieties that flower only during a short period with a precise day length.

For the latitudes of the State of São Paulo (Brazil), the most suitable period for the application of ripening chemicals is between the second half of February and the first half of

March (i.e., during the induction period, to inhibit flowering and to speed ripening). In Guatemala, because of the particular climatic conditions, two applications of products are necessary: ethephon is applied to inhibit flowering and is applied between July and August; then, another application of ripener products is administered to ripen the sugarcane in October.

Research studies involving different sugarcane varieties with flowering habits emphasize that the pith process accompanies flowering, with behavioral differences between varieties for certain technological characteristics (Caputo et al., 2007). Differences in terms of Brix, pol, purity and fiber were observed between the different regions of the stalk compared to the state of the culture (not flowered, in flower, flowered) (Leite et al., 2010).

For the sugarcane agribusiness, flowering is considered to be a waste of the plant's energy, which also ceases the development of the stalk after differentiation of the buds. During flowering, the supply of carbohydrates to the roots decreases to very low levels. In a study of nutrient solutions, it was observed that the roots excrete nitrogenated and potassiated substances into solution during flowering. Various studies have revealed that there is a reduction in the photosynthetic rate during the period of rapid hydrolysis of organic reserves.

With respect to the hormones linked to flowering, auxin, the plant growth hormone, is diminished at the time of flowering. In stalks that bloom, the upper internodes contain the highest fiber content. The percentage of fiber in the top six internodes is 14% higher in flowered sugarcane than in sugarcane that has not flowered.

### 2.5 Plant growth regulators and inhibitors

Plant growth regulators are synthetic substances applied exogenously that have actions similar to known groups of hormones (auxins, gibberellins, cytokinins, retarders, inhibitors and ethylene), while plant hormones are organic, non-nutrient, naturally occurring compounds produced by the plant at low concentrations ($10^{-4}$ M), which promote, inhibit or modify physiological and morphological processes of the plant. Growth inhibitors are natural or synthetic substances that have the capacity to inhibit the growth of the sub-apical meristem.

Currently, most planted areas have already reached a stage that requires high technical skill to achieve economically satisfactory yields. These cultures no longer exhibit nutritional or water limitations and are adequately protected with pesticides. Under these conditions, advanced technology has led to the use of growth regulators, which can often be highly rewarding.

In this context, the use of chemical ripeners in sugarcane, also defined as growth regulators, stands out as an important tool for management. These products are applied to speed the ripening process, promote improvements in the quality of the raw material to be processed, optimize the agribusiness and economic results and aid in the planning of the harvest, as natural ripening in early season can be deficient, even in early varieties.

The ripener paralyzes development, which induces the translocation and storage of sugars, and confers resistance to lodging, which facilitates cutting and reduces losses in the field and the amount of foreign matter transported to the industry. When applied, the ripener is absorbed by the plant and acts by selectively reducing the level of active gibberellin, inducing the plant to temporary reduce its growth rate without affecting the process of photosynthesis and integrity of the apical bud.

Factors such as the period of application of chemical products, doses, the genetic characteristics of the variety and the harvest season of the raw materials are factors that can influence the efficiency of chemical flowering inhibitors and sugarcane ripeners.

### 2.6 Utilization of chemical ripeners

The physiology of sugarcane ripening has been studied for more than 30 years. Natural ripening in early harvest may be poor, even in early varieties. In this context, the use of chemical ripeners stands out as an important tool (Dalley & Richard Junior, 2010). Due to the areas cultivated with sugar cane are extensive, the application of these products is normally done with agricultural aircraft (Fig. 3), but helicopters can also be used for this purpose (Fig. 4).

Fig. 3. Aerial application of ripener using an agricultural aircraft.

Among the chemical products used as ripeners, those that stand out include ethephon (Ethrel, ZAZ and Arvest), a growth regulator; sulfometuron methyl (Curavial), a plant growth regulator from the sulfonylurea chemical group; glyphosate (Roundup, etc.), a growth inhibitor that can cause destruction of the apical bud of the plant and induce lateral growth (which is detrimental to the quality of the raw material); and ethyl-trinexapac (Moddus), which reduces the level of gibberellin without affecting photosynthesis and the integrity of the bud. Other chemicals include fluazifop-p-butyl (Fusilade), maleic hydrazide, paraquat and imazapyr.

Fig. 4. Aerial application of ripener using a helicopter.

### 2.6.1 Ethephon

Ethephon (2-chloroethylphosphonic acid), a chemical from the ethylene-forming group, is a plant growth regulator with systemic properties. Highly soluble in water, it is stable in aqueous solution with a pH $<3.5$ and releases ethylene at higher pH levels. It is sensitive to ultraviolet radiation and is stable up to 75 °C. Ethephon penetrates plant tissues and is progressively translocated, and it then decomposes into ethylene, affecting the growth process (Tomlin, 1994). Its use is justified because the chemical inhibits flowering in sugarcane and increases tillering.

These products that release of ethylene tend to form phosphonic acid. They are maintained at a stable pH of less than or equal to 3.5 (acidic) and lose this stability upon contact with the plant tissue, at which point the pH moves closer to neutrality, releasing gaseous ethylene ($C_2H_4$).

Regarding commercial products, although they have the same phosphonic acid and the technical name of ethephon, they may differ in the pH needed to maintain the stability of the formulation, may include surfactants or other chemical agents and may have concentrations of the active ingredient ranging from 240 g $L^{-1}$ to 720 g $L^{-1}$. Thus, the application rate can vary from 0.67 to 2 L $ha^{-1}$ of commercial product. Castro et al. (2001) reported that different brands of the chemical available on the market did not differ significantly from each other in their capacities to increase the pol% of sugarcane after treatments. Table 1 provides a brief summary of ethephon.

Ethephon has yielded improvements in the technological quality in areas that do not have flowering, and when blooms are present, there is improvement in the production of sugarcane. However, Gururaj Rao et al. (1996) found distinct responses for different varieties of sugarcane when considering production.

For Caputo et al. (2007), ethephon was able to inhibit flowering and to significantly reduce the pith processes of sugarcane varieties, though the magnitude of this reduction varied. Still, it improved the sucrose content in stalks and did not impair the productivity of stalks and sugar.

Studies on the effects of ethephon on sugarcane ripening and productivity emphasized that the product was effective in promoting ripening and increasing sucrose content, allowing the harvest to be anticipated by at least 21 days with a significant reduction in the pith process of the stalk.

With the use of ethephon, an elevated sucrose content from sugarcane has been measured at the beginning of the season, both in experimental and commercial areas. Upon the application of ethephon, there is a temporary shutdown of vegetative growth at the apical meristem. As a result, the produced sugar is stored, resulting in the elevation of the sugar content in the stalks. This change persists for 60 to 90 days, depending on the variety, which is considered to be a long utilization period for the treated cane field. The sugarcane stalks that receive ethephon application always have one or two shorter internodes than normal, indicating that those represent the places of growth at the time of application. As growth intensifies, the sucrose content is reduced, reaching a level that would normally have with no application (i.e., in the middle of the cycle). Samples taken from this period rarely show an economic benefit to the application of ethephon. One of the advantages of its implementation, aside from inhibiting flowering, is to significantly reduce the phenomenon of the pith process, generally resulting in stalks that are denser and that contain greater levels of sucrose.

Another advantage observed with the use of ethephon as a ripener is that it does not damage the sprouting of next sugarcane ratoon; in some cases, a beneficial effect is observed, with increased tillering at the beginning of sprouting after cutting. Silva et al. (2007) observed a stimulating effect on the emergence of tillering up to six months after cutting, although the responses were dependent on the variety.

### 2.6.2 Sulfometuron-methyl

Products of the sulfonylurea chemical group are characterized as potent inhibitors of plant growth, affecting both growth and cell division without interfering directly with mitosis and DNA synthesis. Sulfonylureas inhibit the synthesis of branched-chain amino acids, such as valine, leucine and isoleucine, through the action on the ALS enzyme (acetolactate synthase), which undergoes inhibition of its activity, preventing the synthesis of amino acids from the substrate pyruvate alpha-ketobutyrate. They apparently do not directly block the action of growth promoters (auxins, gibberellins and cytokinins) but strongly stimulate ethylene production due to the stressing effect caused by phytotoxicity. Sulfonylurea molecules originating from foliar or root absorption may be neutral, highly permeable and susceptible to transport in the phloem upon reaching the middle of the cell wall. In alkaline medium, the molecules dissociate in the anionic form, become fixed and systemically move by mass flow through the phloem. These molecules exhibit systemic action, acting in the meristematic regions, affecting growth and inhibiting cell division after absorption by plant leaves. Paralyzed development of the apical meristem causes a reduction in the internodes formed at the time of application. Then, sucrose is stored in the stalk in place of the production of new leaves, which results in a reduction in the rate of the pith process.

The recommended dose of the product to hasten ripening of sugarcane is 15 g $ha^{-1}$ of the active ingredient or 20 g $ha^{-1}$ of commercial product. After application, the treated area can be harvested in 25 to 45 days. Table 1 shows the general characteristics of sulfometuron-methyl.

Studies have reported that sulfometuron-methyl, regarding its potential ripening effect in sugarcane varieties, causes no damage to sugarcane production (t $ha^{-1}$) or the agronomic characteristics of the culture (Silva et al., 2007, Leite et al., 2010). This product does not cause the death of apical buds, and the internodes formed after application resume their normal growth, which allows the culture conditions to be harvested for a longer period. If the harvest of the treated area is late, it does not result in loss or damage to the crop.

The results show consistency in the increase of sugarcane pol, Brix and the reduction index of the pith process (Caputo et al., 2007). Castro et al. (1996) found that the rate of the pith process was reduced from 50 to 60% with the application of sulfometuron-methyl. When sulfometuron-methyl was administered there was an increase of 1.26 in the pol, and ripening occurred 21 days earlier; in addition, treatment induced a decrease in reducing sugars (Caputo et al., 2007, 2008).

The product, when applied to different sugarcane varieties, allows for an improvement in the technological quality of sugarcane: it has a significant determined response regarding gains in pol, increases in purity and reduced organic acid content in the juice and offers a greater possibility for producing higher quality sugar (Fernandes et al., 2002). Organic acids and other undesirable constituents, such as polysaccharides (starches), are responsible for increasing the viscosity of sugar solutions and honey, are precursors to the formation of color (e.g., the ratios of amino acids and reducing sugars) and reduce the exhaustibility of molasses due to the relationship of reducing sugars and ash.

Several studies have identified the influence of the variety on the responses to sulfometuron-methyl (Fernandes et al., 2002, Caputo et al., 2007, 2008). The chemical also does not promote detrimental effects on the sprouting of ratoons following application; although an increase in tillering has been observed up to 180 days after the onset of budding, this condition is not reflected in increased productivity (Silva et al. 2007).

### 2.6.3 Trinexapac-ethyl

Trinexapac-ethyl belongs to the cyclohexanedione chemical group and induces a greater accumulation of sucrose in stalks, facilitating the planning and agro-industrial utilization of sugarcane. This growth regulator inhibits the synthesis of active forms of gibberellic acid, a hormone involved in growth and cell division, which leads to a decrease in plant development and, thus, an accumulation of sucrose.

The recommended dose to promote ripening is between 200 and 300 g $ha^{-1}$ or between 0.8 to 1.2 L pc $ha^{-1}$. It is recommended that the treated area be harvested between 35 and 55 days after application. This and other information regarding trinexapac-ethyl is provided in Table 1.

After application, this product is predominantly absorbed by the leaves and shoots and is translocated to areas of meristematic activity, where it inhibits the elongation of internodes. A shortening of internodes has been observed in different sugarcane varieties, negatively influencing the development of the stalks while improving the technological quality and providing gains of theoretical recoverable sugar in relation to the production of stalks. The suppressive effect on apical elongation leads to formation of shorter internodes, but it does not influence final stalk productivity levels. Still, it is possible to reduce the levels of reducing sugars while also observing increases in Brix concentrations and an increase in juice purity. Other advantages of this product are the control of flowering and the absence of injury in ratoon tillering.

In studies employing trinexapac-ethyl, there were increases in the productivity of sugar and the margin of agricultural contribution, resulting in the improvement of technological quality; additionally, the natural ripening process was anticipated compared to untreated plants (Leite et al., 2008, 2009, 2011).

### 2.6.4 Glyphosate

Glyphosate (N-glycine phosphonomethyl) is currently one of the most popular herbicides in agriculture because of the efficient control it exercises on weeds and its low acute toxicity.

Glyphosate's mechanism of action is quite unique: it is the only herbicide capable of specifically inhibiting the enzyme 5-enolpyruvylshikimate-3-phosphate synthase (EPSPS), which catalyzes the condensation of shikimic acid and phosphate pyruvate, thus preventing the synthesis of three essential amino acids - tryptophan, phenylalanine and tyrosine (Zablotowicz & Reddy, 2004).

Glyphosate is treated as a systemic, non-selective and broad-spectrum herbicide that translocates via the symplast. Its absorption is facilitated by transport proteins of phosphate groups that are present in the membrane. Inhibition of EPSPS leads to accumulation of high levels of shikimate in the vacuoles, which is exacerbated by the loss of feedback control and the unregulated flow of carbon in the pathway. According to Kruse et al. (2000), approximately 35% of dry plant mass is represented by derivatives of the shikimate pathway, and 20% of the carbon fixed by photosynthesis follows this metabolic pathway.

As a ripener, glyphosate has been fairly consistent and effective in speeding sugarcane ripening (Dalley & Richard Junior, 2010) because of two principal reasons:

1. It inhibits sugarcane growth or reduces it by killing the apical bud, or it inhibits the synthesis of indole acetic acid (IAA). Inhibition of stem elongation may also be related to the capacity of auxin to promote ethylene synthesis by increasing the activity of ACC (1-aminocyclopropane-1-carboxylic acid) synthase (Liang et al., 1992). The increase in ethylene may stimulate the senescence process and germination of lateral buds, and the hormonal balance between IAA and ethylene may also lead to the inhibition of stem elongation.
2. It causes stress in sugarcane by inhibiting the synthesis of essential amino acids and proteins. EPSPS is encoded in the nucleus and performs its role in the chloroplast, catalyzing the binding of the compounds shikimate-3-phosphate and phosphoenolpyruvate to produce enolpyruvylshikimate-3-phosphate and inorganic phosphate.

As a ripener, the effective dose of glyphosate varies widely around the world, ranging from applications of 144 to 864 g $ha^{-1}$; this has been due mainly to climatic conditions. In Brazil, the dose is between 144 and 240 g $ha^{-1}$ (i.e., from 0.3 to 0.5 L pc $ha^{-1}$). However, in Guatemala, due to excessive rainfall and high temperatures, which are conditions not conducive to ripening, doses between 0.9 and 1.8 L pc $ha^{-1}$ have been used. Table 1 provides general information on glyphosate.

Results from studies have identified glyphosate as a technical and economical alternative that allows for more flexibility in the harvest period and for managing of the behaviors of different varieties. In the literature, studies have often shown that the application of glyphosate promoted the improvement of the quality of raw material for industry use, increases in sugarcane pol, reduction of the pith process and fiber, reduced juice loss, and reduction in the average number of internodes per stem and in the weight of sugarcane produced.

Studies have also shown that the use of glyphosate as a ripener for sugarcane has promoted increases in recoverable sugar content and sugar production. For distinct varieties of sugarcane, different responses to glyphosate as a ripener have been reported regarding flowering, industrial yield, stalk moisture, Brix, pol and purity.

There are minor differences between the different formulations of glyphosate applied to sugarcane cultures; however, all formulations promoted increases in sucrose content and sugar production compared to controls (Villegas et al., 1993; Bennett & Montes, 2003; Viator et al., 2003).

Fig. 5. Sugarcane stem with side shoots after application of glyphosate.

Some studies have reported two negative effects of glyphosate as a ripener for sugarcane. The first is the elevated rate of side shoots on stems after application (Fig. 5), which leads to lower raw material quality. The second is the detrimental effect on sprouting ratoons after harvest from treated areas, with a reduction of tillers per meter (Fig. 6), which would lead to lower productivity in the next harvest (Leite & Crusciol, 2008). For this reason, glyphosate has been widely used in areas that will be used for the implementation of a new cane field. However, there are also reports that glyphosate did not cause any detrimental effects on sugarcane quality and productivity (Viana et al., 2008). Due to the strong influences that factors such as dose, variety and climatic conditions have on the effectiveness of the product, this issue merits further study.

Fig. 6. Detrimental effect on sprouting ratoons after harvest from glyphosate treated areas.

The Table 1 shows a summary of the main characteristics of ripener glyphosate.

### 2.6.5 Fluazifop-p-butyl, maleic hydrazide, paraquat and imazapyr

There are other products that have been or are currently being used as sugarcane ripeners. These are mainly herbicides, which when used in lower doses, have a stressing action on the plant, promoting ripening. However, with the production of new growth-regulating molecules specific for promoting ripening, these herbicides have been losing importance as ripeners in the management of sugarcane.

Fluazifop-p-butyl is a systemic graminicide herbicide that translocates apoplastically, focusing on the growing points of plants and causing death. However, it may also be used as a ripener in sugarcane when applied at lower doses (from 0.1 to 0.3 L $ha^{-1}$). It is rapidly absorbed in the leaf and causes necrosis; due to its herbicidal action, it kills the apical bud. Therefore, sugarcane should be harvested between 4-6 weeks after application; there is a risk of loss in quality of raw material if this period is exceeded. This product inhibits

| Product | Ethephon | Trinexapac-ethyl | Sulfometuron-methyl | Glyphosate |
|---|---|---|---|---|
| Active Ingredient | 2-chloroethylphosphonic acid | 4-(cyclopropyl-α-hydroxy-methylene)-3,5-dioxo-cyclohexane carboxylic acid ethyl-ester | Methyl-2-[[[[(4,6-dimethyl-2-pyrimidinyl)-amino] carbonyl] amino]sulfonyl]benzoate | N-phosphono methyl glycine |
| Chemical Group | Phosphonic Acid | Cyclohexanedione | Sulfonylurea | Glycine |
| Concentration | 240-720 g $L^{-1}$ | 250 g $L^{-1}$ | 750 g $kg^{-1}$ | 360 g $L^{-1}$ (e.a.) |
| Formulation | CS | CE | GRDA | CS |
| Toxicity Class | II | III | III | IV |
| Vapor Pressure | $1 \times 10^{-7}$ | $1.6 \times 10^{-5}$ | $5.4 \times 10^{-16}$ | $1.8 \times 10^{-7}$ |
| Solubility (ppm) | 1239 | 10.2 | 10 | 10000 |
| Mode of Action | Liberates Ethylene | Inhibits Gibberellin | Inhibits ALS | Inhibits EPSPS |
| Dosage (L or kg $ha^{-1}$) | 0.67-2.0 | 0.8-1.2 | 0.020 | 0.3 – 1.8 |
| Precipitation after Application | 6 hours | 1 hours | 6 hours | 6 hours |
| Harvest (daa) | 45 – 90 | 35 – 55 | 25 – 45 | 25 – 35 |
| Lateral Budding | Yes | Yes | Yes | Yes |
| Inhibition of Flowering | Yes | Yes | Yes | No |
| Stopped Growth | Yes | Yes | Yes | Yes |
| Death of the Apical Bud | Yes | No | No | Yes |
| Varietal Response | Most | All | All | All |
| Period of Use | Beginning of the cycle | All | All | All |
| Rooting | Yes (20%) | Yes (30%) | No | No |
| Germination/ Tillering | Favorable | Favorable | Indifferent | Unfavorable |
| Sprouting of the Ratoon | Favorable | Favorable | Indifferent | Unfavorable |

Table 1. Characteristics of the principal ripeners utilized in sugarcane.

flowering and restricts the volume of the parenchyma without juice (i.e., it decreases the pith process). No negative effect on subsequent ratoon sprouting has been reported, unlike for glyphosate.

Maleic hydrazide (1,2-dihydro-3,6-pyridazinedione) is a plant growth inhibitor that is considered to be a possible ripening agent for sugarcane. This regulator causes a loss of apical dominance in plants. Some monocots showed an increase in sugar content when treated with this ripener. Castro et al. (1985) verified that the application of maleic hydrazide promoted the accumulation of sucrose, though there was a reduction in sugarcane growth. The authors concluded that there was a direct relationship between an increase of the applied dose and an inhibitory effect on plant growth.

Imazapyr is a non-selective systemic herbicide that is absorbed by the leaves and roots. It is rapidly translocated in the xylem and phloem to meristematic regions, where it accumulates. Imazapyr blocks the synthesis of branched-chain amino acids (i.e., valine, leucine and isoleucine) through the inhibition of acetolactate synthase (ALS), which interrupts protein synthesis and leads to interference in DNA synthesis and cell growth. Both effects decrease sugarcane development. Carbohydrates synthesized during photosynthesis after the application of imazapyr are temporarily not used for plant growth; thus, they accumulate in the stalk and increase in concentration. This effect is related to the ripening process because the amount of hydrolysis of sucrose is less than its accumulation in the stem. Lavanholi et al. (2002) observed an increased sugarcane pol with the application of imazapyr, but the authors emphasized that the product did not control flowering.

The application of paraquat, an inhibitor of photosystem I, may or may not affect the quality of industrial stems. Its effect is dose-related when it is used as a desiccant in sugarcane. According to Christoffoletti et al. (1993), the use of paraquat has improved the quality of burn of the cane fields and has yielded raw material with fewer impurities for industrial use. It should be noted that the practice of burning sugarcane fields before harvest is a technique that is being abolished in sugarcane-producing countries, especially in Brazil.

## 3. Conclusion

A supply of raw material of sufficient technological quality to provide economic extraction is one of the greatest needs of the sugarcane industry. For ripening to occur, sugarcane growth must be slowed to accumulate more sucrose. Despite the diversity of genetic material, problems continue to be encountered in the process of providing the sugar industry with raw material throughout the harvest period that contain high levels of sucrose. Sugarcane flowering is also seen as detrimental to the quality of the raw material. Therefore, the application of ripeners and flowering inhibitors is a highly utilized agricultural technique to improve the technological quality of the raw material. The feasibility of using ripeners in the sugarcane production system depends on a number of factors, including climatic, technical and economic variables, and particularly the additional responses that each variety can provide in the practice of this cultivation. Therefore, the producer should consider these factors to find the product that provides the best agricultural, industrial and economic yield.

## 4. References

Alexander, A.G. (1979). *Sugarcane Physiology*. Elsevier, Amsterdam

Araldi, R.; Silva, F.M.L.; Ono, E.O.; Rodrigues, J.D. (2010). Florescimento em cana-de-açúcar. *Ciência Rural*, Vol.40, No..3, (March 2010), pp.694-702, ISSN 0103-8478

Barbieri, V. (1993). *Condicionamento climático da produtividade potencial da cana-de-açúcar (Saccharum* spp.*): Um modelo matemático-fisiológico estatístico de estimativa*. Thesis (PhD in Sciences)- Escola Superior de Agricultura "Luiz de Queiroz", Universidade de São Paulo. Piracicaba.

Bennett, P.G.; Montes, G. (2003). Effect of glyphosate formulation on sugarcane ripening. *Sugar Journal*, Vol. 66, No.1, (January, 2003), pp. 22, ISSN 0039-4734

Botha, F.C.; Black, K.G. (2000). Sucrose phosphate synthase and sucrose synthase activity during maturation of internodal tissue in sugarcane. *Australian Journal of Plant Physiology*, Vol.27, No.1, pp.81-85, ISSN 0310-7841

Caputo, M.M. (2006). *Indução da maturação por produtos químicos e sua conseqüência na qualidade tecnológica de diferentes genótipos de cana-de-açúcar*. Dissertation (Master in Science in Crop Production). Escola Superior de Agricultura "Luiz de Queiroz", Universidade de São Paulo, Piracicaba.

Caputo, M.M.; Beauclair, E.G.F.; Silva, M.A.; Piedade, S.M.S. (2008). Resposta de genótipos de cana-de-açúcar à aplicação de indutores de maturação. *Bragantia*, Vol.67, No.1, (January 2008), pp.15-23. ISSN 0006-8705

Caputo, M.M.; Silva, M.A.; Beauclair, E.G.F.; Gava, G.J.C. (2007). Acúmulo de sacarose, produtividade e florescimento de cana-de-açúcar sob reguladores vegetais. *Interciencia*, Vol.32, No. 12, (December 2007), pp.834-840, ISSN 0378-1844

Castro, P.R.C. (1999). Maturadores químicos em cana-de-açúcar, *Proceedings of SECAPI 1999 4th Semana da Cana-de-açúcar de Piracicaba*, pp.12-16, Piracicaba, São Paulo, Brazil, 1999.

Castro, P.R.C. (2002). Efeitos da luminosidade e da temperatura na fotossíntese e produção e acúmulo de sacarose e amido na cana-de-açúcar. *STAB. Açúcar, Álcool e Subprodutos*, Vol.20, No. 5, (May 2002), pp. 32-33, ISSN 0102-1214

Castro, P.R.C.; Appezzato, B.; Gonçalves, M.B. (1985). Ação de hidrazida maleica e etefon no crescimento da cana-de-açúcar. *Anais da E. S. A. "Luiz de Queiroz"*, Vol.42, No.2, (April 1985), pp. 391-399, ISSN 0071-1276

Castro, P.R.C.; Miyasaki, J.M.; Bemardi, M.; Marengo, D.; Nogueira, M.C.S. (2001). Efeito do ethephon na maturação e produtividade da cana-de-açúcar. *Revista de Agricultura*, Vol. 76, No. 2, (May 2001), pp. 277-290, ISSN 0034-7655

Castro, P.R.C.; Oliveira, D.A.; Panini, E.L. (1996). Ação do sulfometuron metil como maturador da cana-de-açúcar, *Proceedings of 6th Congresso Nacional da Sociedade dos Técnicos Açucareiros de Alcooleiros do Brasil*, p. 363-369, Maceió, Alagoas, Brazil, November 24-29, 1996

Christoffoletti, P.J.; Sacomano, J.; Soffner, R.; Cattaneo, S. (1993). Avaliação do uso de Paraquat como dessecante na cultura da cana-de-açúcar em condições de pré-colheita. *Revista de Agricultura*, Vol. 68, No. 3, (September 1993), pp. 271-286, ISSN 0034-7655

Dalley, C.D., Richard Junior, E.P. (2010). Herbicides as ripeners for sugarcane. *Weed Science*. Vol.58, No.3, (July 2010), pp. 329-333, ISSN 1550-2759

Deuber, R. (1988). Maturação da cana-de-açúcar na região sudeste do Brasil, *Proceedings of 4th* Seminário de Tecnologia COPERSUCAR 1988, pp. 33-40, Piracicaba, São Paulo, Brazil

Ebrahim, M.K.; Zingsheim, O.; El-Shourbagy, M.N.; Moore, P.H.; Komor, E. (1998). Growth and sugar storage in sugarcane grown at temperatures below and above optimum. *Journal of Plant Physiology*, Vol. 153, No. 5/6, (May 1998), pp. 593-602, ISSN 0176-1617

Fernandes, A.C. (1982). Refratômetro de campo. *Boletim Técnico Coopersucar*, Vol.19, (July 1982), pp.5-12.

Fernandes, A.C.; Stupiello, J.P.; Uchoa, P.E. de A. (2002). Utilização do Curavial para melhoria da qualidade da cana-de-açúcar. *STAB. Açúcar, Álcool e Subprodutos*, Vol.20, No.4, (March 2002), p.43-46, ISSN 0102-1214

Glover, J. (1972). Practical and theoretical assessments of sugarcane yield potential in Natal, *Proceedings of 46th* South African Sugarcane Technologists Association, 1972, pp.138-141, Natal, South Africa.

Gururaja Rao, P.N.; Singh, S.; Mohan Naidu, K. (1996). Flowering suppression by ethephon in sugarcane and its effect on yield and juice quality. *Indian Journal of Plant Physiology*, Vol.1, No. 4, (December 1996), pp. 307-309, ISSN 0974-0252

Keating, B.P.; Robertson, M.J.; Muchow, R.C.; Huth, N.I. (1999). Modelling sugarcane production systems. I. Development and performace of the sugarcane module. *Field Crops Research*, Vol. 61, No.3, (May 1999), pp.253-271, ISSN 0378-4290

Kruse, D.N.; Trezzi, M.M.; Vidal, R.A. (2000) Herbicidas inibidores da EPSPS: Revisão de literatura. *Revista Brasileira de Herbicidas*, Vol. 1, No. 2, (May 2000), pp. 139-144, ISSN 1517-9443

Lavanholi, M.G.D.P.; Casagrande, A.A.; Oliveira, L.A.F. de; Fernandes, G.A.; Rosa, R.F. da (2002). Aplicação de etefon e imazapyr em cana-de-açúcar em diferentes épocas e sua influência no florescimento, acidez do coldo e teores de açúcares nos colmos - variedade SP70-1143. *STAB. Açúcar, Álcool e Subprodutos*, Vol.20, No. 5, (May 2002), pp. 42-45, ISSN 0102-1214

Legendre, B.L. (1975). Ripening of sugarcane: effects of sunlight, temperature, and rainfall. *Crop Science*, Vol.15, No.3, (May 1975), pp. 349-352, ISSN 1435-0653

Leite, G.H.P.; Crusciol, C.A.C. (2008). Reguladores vegetais no desenvolvimento e produtividade da cana-de-açúcar. *Pesquisa Agropecuária Brasileira*, Vol.43, No.8, (August 2008), pp.995-1001, ISSN 1678-3921

Leite, G.H.P.; Crusciol, C.A.C.; Lima, G.P.P.; Silva, M.A. (2009). Reguladores vegetais e atividade de invertases em cana-de-açúcar em meio de safra. *Ciência Rural*, Vol.39, No.3, (May 2009), pp.718-725, ISSN 0103-8478

Leite, G.H.P.; Crusciol, C.A.C.; Silva, M.A. (2011). Desenvolvimento e produtividade da cana-de-açúcar após aplicação de reguladores vegetais em meio de safra. *Semina: Ciências Agrárias*, Vol.32, No.1, (January 2011), pp.129-138, ISSN 1679-0359

Leite, G.H.P.; Crusciol, C.A.C.; Silva, M.A.; Venturini Filho, W.G. (2008). Reguladores vegetais e qualidade tecnológica da cana-de-açúcar em meio de safra. *Ciência e Agrotecnologia*, Vol.32, No.6, (November 2008), pp.1843-1850, ISSN 1413-7054

Leite, G.H.P.; Crusciol, C.A.C.; Siqueira, G.F.; Silva, M.A. (2010). Qualidade tecnológica em diferentes porções do colmo e produtividade da cana-de-açúcar sob efeito de maturadores. *Bragantia*, Vol.69, No.4, (October 2010), pp.861-870, ISSN 0006-8705

Liang, X.; Steffen, A.; Keller, J.A.; Shen, N.F.; Theologis, A. (1992). The 1-aminociclopropane-1-carboxilate synthase genmne family of *Arabidopsis thaliana*. *Proceedings of the National Academy of Science of the United States of America*. Vol.89, No.22, (November 1992), pp.11046-11050, ISSN 1091-6490

Lingle, S.E. (1997). Seasonal internode development and sugar metabolism in sugarcane. *Crop Science*, Vol.37, No. 4, (July 1997), pp.1222-1227, ISSN 1435-0653

Lingle, S.E. (1999). Sugar metabolism during growth and development in sugarcane internodes. *Crop Science*, Vol.39, No. 2, (March 1999), pp.480-486, ISSN 1435-0653

Melotto, E.; Castro, P.R.C.; Godoy, O.P.; Câmara, G.M.S.; Stupiello, J.P.; Iemma, A.F. (1987). Desenvolvimento da cana-de-açúcar cultivar NA56-79 proveniente da propagação de colmos tratados com ethephon. *Anais da E. S. A. "Luiz de Queiroz"*, Vol.44, No.1, (January 1987), pp. 657-676, ISSN 0071-1276

Moore, P.H. (1995). Temporal and spatial regulation of sucrose accumulation in the sugarcane stem. *Australian Journal of Plant Physiology*, Vol.22, (February 1995), pp. 661-679, ISSN 0310-7841

Rohwer, J.M.; Botha, F.C. (2001). Analysis of sucrose accumulation in the sugar cane culm on the basis of in vitro kinetic data. *Biochemical Journal*, Vol.358, No. 2, (September 2001), pp.437-445, ISSN 0264-6021

Rose, S.; Botha, F.C. (2000). Distribution patterns of neutral invertase and sugar content in sugarcane internodal tissues. *Plant Physiology and Biochemistry*, Vol.38, No.11, (November 2000), pp.819-824, ISSN 0981-9428

Scarpari, M.S.; Beauclair, E.G.F. (2004). Sugarcane maturity estimation through edaphic-climatic parameters. *Scientia Agricola*, Vol.61, No.5, (September 2004), pp.486-491, ISSN 0103-9016

Silva, M.A.; Gava, G.J.C; Caputo, M.M.; Pincelli, R.P.; Jerônimo, E.M.; Cruz, J.C.S. (2007). Uso de reguladores de crescimento como potencializadores do perfilhamento e da produtividade em cana-soca. *Bragantia*, Vol.66, No.4, (October 2007), pp.545-552, ISSN 0006-8705

Siqueira, G. F. (2009). *Eficácia da mistura de glifosato a outros maturadores na cana-de-açúcar (Saccharum spp.)*. Dissertation (Master in Science in Agriculture). Faculdade de Ciências Agronômicas, Universidade Estadual Paulista, Botucatu.

Suzuki, J. (1983). *Biossíntese e acúmulo de sacarose em cana-de-açúcar (Saccharum* spp.*): Influência do íon Potássio durante diferentes estádios de crescimento em solução nutritiva*. Thesis (PhD in Soil and Plant Nutrition). Escola Superior de Agricultura "Luiz de Queiroz", Universidade de São Paulo, Piracicaba.

Terauchi, T.; Matsuoka, M.; Kobayashi, M.; Nakano, H. (2000). Activity of sucrose phosphate synthase in relation to sucrose concentration in sugarcane internodes. *Japanese Journal of Tropical Agriculture*, Vol.44, No.3, (July 2000), pp.141-151, ISSN 0021-5260

Tomlin, C. (1994). *The pesticide manual*. (10th. Edition), Blackwell Scientific Publications, Cambridge

Viana, R.S.; Silva, P.H.; Mutton, M.A.; Mutton, M.J.R.; Guimarães, E.R.; Bento, M. (2008). Efeito da aplicação de maturadores químicos na cultura da cana-de-açúcar (*Saccharum* spp.) variedade SP81-3250. *Acta Scientiarum Agronomy*. Vol.30, No.1, (January 2008), pp. 65-71, ISSN 1807-8621

Viator, B.J.; Viator, C.; Jackson, W.; Waguespack, H.; Richard Junior., E.P. (2003). Evaluation of potassium-based ripeners as an alternative to glyphosate and the effects of 2,4-D on herbicidal cane ripening. *Sugar Journal*, Vol.66, No.1, pp. 21, ISSN 0039-4734

Villegas, F.T.; Torres, J.S.A. (1993). Efecto del Roundup usado como madurante en la producción de caña de azúcar. *International Sugar Journal*, Vol.95, No.1130, pp. 59-64, ISSN 0020-8841

Vorster, D.J.; Botha, F.C. (1999). Sugarcane internodal invertases and tissue maturity. *Journal of Plant Physiology*, Vol.155, No.4/5, (April 1999), pp.470-476, ISSN 0176-1617

Zablotowicz, R.M.; Reddy, K.N. (2004). Impact of glyphosate on the *Bradyrzobium japonicum* symbiosis with glyphosate-resistant transgenic soybean: a minirevew. *Journal of Environmental Quality*, Vol.33, No.3, (May 2004), pp. 825-831, ISSN 1537-2537

Zhu, Y.J.; Komor, E.; Moore, P.H. (1997). Sucrose accumulation in the sugarcane stem is regulated by the difference between the activities of soluble acid invertase and sucrose phosphate synthase. *Plant Physiology*, Vol.115, No.2, (October 1997), pp.609-616, ISSN 1532-2548

# 3

# Managing Cover Crops for Conservation Purposes in the Fraser River Delta, British Columbia

Jude J. O. Odhiambo[1*], Wayne D. Temple[2] and Arthur A. Bomke[2]
[1]*University of Venda, Thohoyandou*
[2]*University of British Columbia*
*Faculty of Land and Food Systems, Vancouver, B.C.*
[1]*South Africa*
[2]*Canada*

## 1. Introduction

The farmland of the Fraser River delta is some of Canada's most productive because of a unique combination of climate and soil. Much of the cultivated farmland or soil-based agriculture of the Fraser River delta is located within the Municipality of Delta which is part of the highly populated and urbanized environment of the Greater Vancouver Regional District (GVRD). On average, the farmlands of Delta receive about 1000 mm of precipitation per annum, of which approximately 75% falls between November and March (Bomke et al., 1994); and the longest period of frost-free days in Canada, extending from April 15 to October 21. In Delta, the soils are inherently fertile, heavy in texture and deep (Luttmerding 1981); consequently the soil has a good water storage capacity and a potential to sustain crop production on a year-round basis. Approximately 8000 hectares are being farmed in Delta (Klohn et al., 1992). Results from this work should be relevant on similar soils throughout the Georgia Basin of British Columbia and Washington as well as in parts of Western Oregon.

Present day farming in Delta is far below its potential crop productivity (Klohn et al., 1992). The reason for this is related to a number of soil factors, but most notably inadequate sub-surface drainage, deep soil compaction and declining soil organic matter levels. In the past thirty years, about half the farm acreage in Delta has shifted from a mixed farming practice to cultivated vegetable production. At present three quarters of the farmers produce vegetables (i.e. green beans, peas, corn, cabbage and potatoes) with small areas of berries and cereal crops. As a consequence, there have been very significant losses in soil organic matter and associated problems with losses in soil surface structure and with deep soil compaction from working soils that are too wet. Many of the Delta growers view the above as a serious limitation to crop productivity. Some of these degraded fields exhibit uneven seed germination or crop establishment and even crop failures during wet or very dry

* Corresponding Author

growing seasons. Optimum soil and crop management practices must now be identified which will first reclaim and then sustain these soils at their former or improved capability for crop production. The absence of any significant livestock industry in Delta has greatly limited the availability of soil organic matter inputs such as barnyard manures. Therefore, growing cover crops in Delta as a green manure can help maintain many conservation objectives, such as:

- providing much needed organic matter to the soil;
- protecting and improving the soil surface structure;
- sustaining or increasing earthworm populations;
- reducing the incidence of weeds; and
- providing forage for waterfowl.

The cash crop rotations, soil characteristics, moderate climate and the presence of high numbers of grazing waterfowl combine to give Delta unique constraints for selection of winter cover crops for conservation objectives (Temple et al., 1991). Goals such as winter soil cover and green manuring for soil structure improvement are high priorities while nitrogen conservation is less an environmental imperative than in areas with sensitive groundwater resources.

In this study, a wide range of different cover crop species were screened over a five-year period (beginning in 1991) under conditions of high winter rainfall, poor soil surface structure and intensive waterfowl grazing. Other companion studies to this investigation include the effects of cereal and mixed legume cover crops on soil N cycling (Odhiambo & Bomke, 2000) and the effects of the cover crops on soil environmental parameters such as labile polysaccharides, aggregate mean weight diameter, earthworm numbers and infiltration rates (Hermawan, 1995; Hermawan & Bomke, 1996 ;Liu, 1995). The objective of this study was to screen a wide range of over-winter cover crops with respect to their growth, development, soil surface protection, green manure production and residual soil nitrate (N-uptake) conservation.

## 2. Materials and methods

### 2.1 Field sites

In the late summer and fall periods of 1991 through to 1995, cover crop screening trials were established in cooperation with farmers on Westham Island, located within the municipality of Delta, British Columbia. The soils were developed on moderately fine textured deltaic deposits and were mapped as either Westham or Crescent series (Luttmerding, 1981). At each site various cover crops were planted (see Table 1 for planting dates) in the third week of August (early seeding) and third week of September (late seeding). Cover crops were seeded in a randomized complete block design with four replicates. Early and late seeded treatments were conducted on adjacent, but separate experimental areas. The first planting date was to coincide with a cover crop planting after the early harvest of cash crops (i.e. field peas (*Pisum sativum* L.), early harvested potatoes (*Solanum tuberosum* L.) and green beans (*Phaseolus vulgaris* L.), cabbage (*Brassica pekinensis* L.) and corn (*Zea mays* L.). The second planting date was to coincide with the late harvest of cash crops (i.e. green beans, corn, cabbage and potatoes). Soils were lightly disked just prior to planting to incorporate previous crop residues and weeds.

| Year | Previous crop | Planting dates: Early seeding | Late seeding | Harvest dates: Fall harvest | Spring harvest |
|---|---|---|---|---|---|
| 1991/92 | Potato | 17-Aug | 18-Sep | 1&15-Nov[1] | 8-Apr |
| 1992/93 | Green Bean | 25-Aug | 22-Sep | 7-Nov | 30-Apr |
| 1993/94 | Cabbage | 24-Aug | 22-Sep | 22-Nov | 22-Apr |
| 1994/95 | Green Bean | 23-Aug | 20-Sep | 21-Nov | 27-Apr |
| 1995/96 | Potato | 23-Aug | 23-Sep | 12-Nov | 7-May |

[1]The second late seeding date was harvested two weeks later in fall 1991.

Table 1. Site information for cover crop screening trials

Cover crops were seeded with a Vicon air-seeder at 0.10m row spacings in 3m x 10m plots. Cover crops included fall rye (*Secale cereale* L.; three cultivars), winter wheat (*Triticum aestivum* L.; two cultivars), annual ryegrass (*Lolium multiflorum* Lam.), winter triticale (*Triticum durum x Secale cereale* hybrid; two cultivars), winter barley (*Hordeum vulgare* L.), spring barley (*Hordeum vulgare* L.; three cultivars), spring oats (*Avena sativa* L.), spring wheat (*Triticum aestivum* L.), forage rape (*Brassica napus* L.), forage kale (*Brassica oleracea var. acephala* L.), crimson clover (*Trifolium incarnatum* L.), alsike clover (*Trifolium hybridum* L.) and red clover (*Trifolium pratense* L.). Sudan grass (*Sorghum sudanense* L.), Austrian winter peas (*Pisum sativum* L.) and buckwheat (*Fagopyrum esculentum* L.) were also included in the first year. Refer to Table 3 for cultivars used and planting densities.

### 2.2 Plant sampling methods

The cover crops were hand clipped (see Table 1 for harvest dates) using 0.25 $m^2$ quadrats in November when some of the crops began to show signs of die-back (frost damage) and when fields were beginning to become too soft to walk on because of heavy November rains. Plots were sampled again in April just prior to the preparation of seedbeds for planting summer crops. Sub samples were oven-dried for 72 h at 65°C and all biomass yields are given on a dry weight basis. A sub sample was ground (1mm sieve), digested according to the Parkinson and Allen (1975) method and analyzed for total N using a Technicon Auto analyzer (Technicon, 1974). In November, cover crops were measured or visually rated for cover (point method), early establishment, cold tolerance and height. Weed control assessments were made in April, just prior to plow down, and based upon % cover.

### 2.3 Statistical analysis

Seeding of early and late cover crop planting dates was analyzed separately. The analyses were conducted using Analysis of Variance (ANOVA) and General Linear Models (GLM) procedure of windows SAS Version 9.1 (SAS Institute, Cary, NC). For plant biomass and N-uptake, mean values were compared using Duncan's multiple range test following a significant F value at P=0.05.

## 3. Results and discussion

### 3.1 Climate and weather conditions

Table 2 presents the 30-year normal and monthly temperatures and precipitation for the years that the screening trials were performed. Relative to the normal precipitation for the

November to April cover crop growing season, November 1991-92 was about average, January and April were wetter months, while December, February and March were drier. November and March of 1992-93 were about average; April was wetter, while December, January and February were drier months than normal. The 1993-94 field season's (El Niño year) November and January were relatively dry, while December, February, March and April were about average. November and January of 1994-95 were wetter months, while December, February, March and April were about average. The 1995-96 field season's November, December and April were wetter; January and February were about average and March was a drier month. With the exception of the warm December of 1991 and colder than normal January of 1993, monthly temperatures did not appear to vary enough from the 30-year normal temperatures to greatly influence crop performance.

| | | | Mean Monthly Temperature (oC) | | | | |
|---|---|---|---|---|---|---|---|
| Month | 30yr. Normal | 1991 | 1992 | 1993 | 1994 | 1995 | 1996 |
| Jan | 3.0 | 1.6 | 5.8 | -0.4 | 6.3 | 4.5 | 2.9 |
| Feb | 4.7 | 7.1 | 6.6 | 3.5 | 3.7 | 5.3 | 4.3 |
| Mar | 6.3 | 5.4 | 8.5 | 7.4 | 7.2 | 7.1 | 6.8 |
| Apr | 8.8 | 8.8 | 10.6 | 10.0 | 10.9 | 9.6 | 10.4 |
| May | 12.1 | 12.3 | 13.5 | 14.7 | 13.8 | 14.2 | 11.7 |
| Jun | 15.2 | 14.6 | 17.2 | 15.8 | 15.0 | 16.7 | 15.2 |
| Jul | 17.2 | 17.8 | 18.4 | 16.4 | 18.5 | 18.5 | 18.2 |
| Aug | 17.4 | 18.0 | 17.8 | 17.6 | 18.5 | 16.5 | 18.1 |
| Sep | 14.3 | 14.9 | 13.9 | 14.8 | 15.7 | 16.6 | 13.7 |
| Oct | 10.0 | 9.2 | 11.3 | 11.4 | 10.2 | 10.4 | 9.6 |
| Nov | 6.0 | 7.1 | 6.4 | 4.5 | 5.0 | 8.0 | 5.0 |
| Dec | 3.5 | 5.6 | 1.9 | 4.5 | 4.4 | 4.7 | 1.4 |
| | | | Total Monthly Precipitation (mm) | | | | |
| Month | 30yr. Normal | 1991 | 1992 | 1993 | 1994 | 1995 | 1996 |
| Jan | 149.8 | 156.8 | 281.8 | 103.4 | 112.5 | 164.4 | 160.4 |
| Feb | 123.6 | 143.3 | 87.8 | 11.4 | 108.2 | 140.1 | 110.8 |
| Mar | 108.8 | 106.4 | 25.9 | 115.2 | 103.2 | 110.9 | 70.3 |
| Apr | 75.4 | 111.3 | 126.2 | 126.9 | 65.0 | 54.4 | 171.5 |
| May | 61.7 | 60.2 | 15.8 | 100.8 | 39.6 | 27.2 | 72.3 |
| Jun | 45.7 | 53.6 | 96.4 | 72.2 | 70.5 | 46.0 | 13.6 |
| Jul | 36.1 | 33.6 | 27.1 | 34.3 | 27.4 | 43.4 | 17.2 |
| Aug | 38.1 | 170.0 | 23.2 | 19.0 | 18.0 | 69.5 | 33.8 |
| Sep | 64.4 | 8.8 | 48.2 | 2.1 | 65.6 | 15.2 | 69.8 |
| Oct | 115.3 | 27.1 | 109.1 | 73.1 | 113.0 | 141.4 | 249.6 |
| Nov | 169.9 | 192.4 | 168.3 | 63.1 | 205.2 | 251.7 | 214.5 |
| Dec | 178.5 | 95.8 | 117.8 | 162.3 | 189.1 | 221.0 | 280.6 |
| TOTALS | 1167 | 1159 | 1128 | 884 | 1117 | 1285 | 1464 |

Table 2. 30 year normal and monthly temperature and precipitation for study area as recorded at Vancouver International Airport (source: Environment Canada).

### 3.2 Growth, development and soil surface protection

Sudan grass barely emerged and died very quickly after the first frost. Buckwheat established well, but produced very little biomass and it too died very quickly after the first frost. Winter peas did not establish well and were subject to high disease pressure. The sudan grass, buckwheat and winter peas were subsequently covered by weeds. None of these three cover crops warranted sampling and are not recommended for over-winter cover cropping in this region. Hence they are not considered under results and discussions. Table 3 presents the cover crop biomass yields and Table 4 summarizes some of the field observations with respect to establishment, % cover and cold tolerance, November sampling height of crops, and weed control for each of the most commonly used cover crops in the screening trials. With the exception of triticale, red clover and alsike clover, all of the cover crops seeded in the third week of August achieved 80 -100% cover. Spring cereals (wheat, barley and oats) fall rye, winter barley, winter wheat, annual ryegrass and forage rape all gave adequate winter cover. Given reasonable conditions for emergence and moderate drainage, the crimson clover seeded in August produced satisfactory cover. Complete soil cover often occurred in the absence of an extensive cover crop canopy as weeds, especially chickweed (*Stellaria media* L.) filled in bare areas. Over-winter weed control was usually directly related to the cover crops' ability to rapidly establish soil cover. It should also be noted that early seeded cereals, in particular spring and winter barley, triticale and spring oats, were frequently infected with fungal diseases including brown leaf rust, septoria and powdery mildew. These could increase the need for fungicide sprays when over-wintering cereals are used as cover crops in crop rotations with a high proportion of cereals as cash crops.

The use of similar crop species as winter and summer cash crops can provide an undesirable "green bridge" enabling the survival of pests common to both. This concern could also apply to the use of *Brassica* cover crops, forage rape and kale, when the crop rotation includes cabbage, cauliflower etc. Poor establishment of Austrian winter peas in the single trial in which it was planted may indicate a similar negative relationship to crop rotations that include processing peas.

The accumulation of cover crop biomass is directly related to its ability to improve soil structure and to take up and conserve residual nitrogen prior to the winter rainy period. In this regard, the cover crops screened in our trials fall into three categories: 1) frost sensitive spring cereals; 2) true winter cereals; and 3) species which grow rapidly in the fall, yet exhibit good cold tolerance. With the exception of frost sensitive cover crops, such as spring barley and oats, much of the biomass accumulation occurs in the spring regardless of the planting date. Examples of the latter group are 'Aubade' annual ryegrass, 'Danko' fall rye and the spring wheat cultivar 'Max'. These crops have proven to be effective in relatively cold growing conditions; continuing growth even after frosts and during cool temperatures during the winter. Therefore these cover crops are recommended for late planting dates.

The frost-sensitive spring cereals were barley cultivars 'Virden', 'Harrington' and oat cultivars 'Winchester' and 'Cascade'. Winter-kill of spring cereals, which resulted in good straw mulch, only occurred with crops planted early. Late planted spring cereals did not die-back, became diseased and provided little soil surface protection. There were a few differences among the spring cereals; for example, early-planted spring oats produced less biomass than the spring barley cultivars by the end of fall 1992. The differentiation among spring barley cultivars is small enough that farmers requiring a cover crop that will winter-kill can choose from any of the above cultivars depending on price and availability. Spring barley is preferred over oats because we observed good control of spring weeds, likely due to allelopathic effects. 'Virden' barley, because of its higher grain yields when grown locally and its straw strength should have lower costs of seed production and may be a good cultivar to emphasize.

| | | Fall & Spring dry matter yields (t/ha) for early & late planting dates | | | | | | | |
|---|---|---|---|---|---|---|---|---|---|
| Year: | | 1991/92 | | | | 1992/93 | | | |
| Harvest dates[1]: | | Fall | | Spring | | Fall | | Spring | |
| Planting dates[1]: | | Early | Late | Early | Late | Early | Late | Early | Late |
| Cover Crop; *Variety* | Planting rate (kg/ha) | | | | | | | | |
| Fall rye; *Ladner Common* | 100 | 2.93 efg** | 0.65 c | 6.40 cd | 5.91 bc | 1.37 bcd | 1.20 a | 6.06 ab | 3.42 bcd |
| Fall rye; *Danko* | 100 | 3.61 def | 0.94 abc | 5.19 cd | 3.98 cd | 1.68 ab | 0.78 b | 6.37 ab | 3.60 bc |
| Fall rye; *Kodiac* | 100 | ——* | —— | —— | —— | —— | | —— | —— |
| Winter wheat; *Monopol* | 100 | 2.71 fg | 0.68 bc | 7.44 bc | 6.79 b | 1.07 cd | 0.32 de | 6.80 a | 2.17 def |
| Winter wheat; *Fundelea* | 100 | —— | 0.58 c | —— | 7.89 ab | —— | —— | —— | —— |
| Annual ryegrass; *Aubade* | 25 | 4.01 cde | 0.57 c | 8.89 ab | 6.84 b | 1.43 bc | 0.32 de | 5.71 abc | 4.54 ab |
| Winter triticale; *Wintri* | 100 | 3.39 efg | 0.67 bc | 7.24 bc | 3.51 d | 0.94 d | NH | 4.24 cd | 2.09 ef |
| Winter triticale; *Pika* | 100 | —— | —— | —— | —— | —— | —— | —— | —— |
| Winter barley; *Elmira* | 100 | 3.13 efg | 0.97 abc | 6.42 cd | 4.49 cd | 1.38 bcd | 0.25 e | 4.80 bcd | 2.46 cdef |
| Spring barley; *Winchester* | 100 | 5.58 ab | 0.83 abc | WK | 3.51 d | 1.75 ab | 0.50 cd | WK | WK |
| Spring barley; *Harrington* | 100 | 5.78 a | 1.08 abc | WK | WK | 1.69 ab | 0.56 bcd | WK | WK |
| Spring barley; *Virden* | 100 | 5.58 ab | 1.28 ab | WK | WK | 2.08 a | 0.57 bc | WK | WK |
| Spring oats; *Cascadia* | 100 | 5.36 ab | 1.11 abc | WK | 5.93 bc | 1.16 cd | NH | WK | WK |
| Spring wheat; *Max* | 100 | 5.01 abc | 1.14 abc | WK | 9.11 a | 1.09 cd | 0.25 e | 5.27 abcd | 4.85 a |
| Forage rape; *Liratop* | 10 | 4.52 bcd | 1.43 a | 9.97 a | 6.73 b | 1.77 ab | 0.49 cde | 4.91 bcd | 3.00 cde |
| Forage kale; *Premier* | 8 | —— | —— | —— | —— | —— | 0.45 cde | —— | 1.56 f |
| Crimson clover; *Common* | 18/12** | 2.41 g | —— | 4.31 de | —— | NH | —— | 3.75 d | —— |
| Alsike clover; *Common* | 7 | 0.73 h | —— | 2.10 f | —— | —— | —— | —— | —— |
| Red clover; *Pacific Red* | 12 | 1.28 h | —— | 2.74 ef | —— | NH | —— | 0.99 e | —— |
| p-values: | | 0.0001 | 0.0256 | 0.0001 | 0.0001 | 0.0001 | 0.0001 | 0.0001 | 0.0001 |
| C.V.: | | 19.4 | 30.1 | 23.4 | 21.9 | 20.4 | 28.9 | 21.8 | 25.8 |

[1]See Table 1 for planting and harvest dates.

*Cover crop not planted in that year or on that date.

**Means in the same column followed by the same letter are not significantly different; alpha=0.05.

#WK = winter kill; NH = insufficient sample to harvest, height less than 5 cm.

**Seeding rate; 18kg/ha in 1991/92 & 1992/93; 12 kg/ha in 1993/94 & 1994/95.

Table 3. Cover crop screening trials: Yearly harvest above ground dry matter yields

| Fall & Spring dry matter yields (t/ha) for early & late planting dates | | | | | | | | | | | |
|---|---|---|---|---|---|---|---|---|---|---|---|
| 1993/94 | | | | 1994/95 | | | | 1995/96 | | | |
| Fall | | Spring | | Fall | | Spring | | Fall | | Spring | |
| Early | Late | Early | Late | Early | Late | Early | Late | Early | Late | Early | Late |
| ------ | ------ | ------ | ------ | ------ | ------ | ------ | ------ | ------ | ------ | ------ | ------ |
| 1.30 def | 0.59 c | 2.73 c | 2.81 d | 2.38 bc | 1.57 | 5.67 b | 4.26 a | ------ | ------ | ------ | ------ |
| ------ | ------ | ------ | ------ | ------ | ------ | ------ | ------ | 4.17 a | NH | 6.46 | 5.51 b |
| 1.55 cdef | 0.43 d | 5.79 a | 4.85 b | 1.80 cd | NH | 9.90 a | 5.63 a | 2.87 b | NH | 9.43 | 4.76 bc |
| ------ | ------ | ------ | ------ | ------ | ------ | ------ | ------ | ------ | ------ | ------ | ------ |
| 2.08 bcd | 0.28 e | 5.11 ab | 4.24 bc | 0.97 d | NH | 8.12 a | 5.18 a | 4.70 a | NH | 7.40 | 10.8 a |
| ------ | ------ | ------ | ------ | ------ | ------ | ------ | ------ | ------ | ------ | ------ | ------ |
| 1.22 ef | 0.45 cd | 3.26 bc | 3.31 cd | ------ | ------ | ------ | ------ | ------ | ------ | ------ | ------ |
| ------ | ------ | ------ | ------ | ------ | ------ | ------ | ------ | ------ | ------ | ------ | ------ |
| ------ | ------ | ------ | ------ | ------ | ------ | ------ | ------ | ------ | ------ | ------ | ------ |
| ------ | ------ | ------ | ------ | ------ | ------ | ------ | ------ | ------ | ------ | ------ | ------ |
| 2.74 ab | 0.75 b | WK | WK | 3.68 a | 1.65 | WK | WK | 2.91 b | NH | WK | 3.92 c |
| 1.99 bcde | 0.43 d | WK | WK | ------ | ------ | ------ | ------ | ------ | ------ | ------ | ------ |
| 3.27 a | 0.50 cd | WK | 6.71 a | 3.12 ab | 1.04 | WK | 5.54 a | 4.55 a | NH | WK | 5.45 b |
| 2.26 bc | 1.04 a | 2.17 c | 4.98 b | ------ | ------ | ------ | ------ | ------ | ------ | ------ | ------ |
| ------ | ------ | ------ | ------ | ------ | ------ | ------ | ------ | ------ | ------ | ------ | ------ |
| 1.09 f | NH | 3.53 bc | 1.06 e | 0.97 d | NH | 4.93 b | 1.90 b | ------ | ------ | ------ | ------ |
| ------ | ------ | ------ | ------ | ------ | ------ | ------ | ------ | ------ | ------ | ------ | ------ |
| ------ | ------ | ------ | ------ | ------ | ------ | ------ | ------ | ------ | ------ | ------ | ------ |
| 0.0001 | 0.0001 | 0.0151 | 0.0001 | 0.0001 | 0.2196 | 0.0011 | 0.0005 | 0.0005 | ------ | 0.1450 | 0.0001 |
| 26.5 | 17.1 | 36.8 | 19.8 | 28.6 | 33.3 | 17.4 | 22.6 | 13.9 | ------ | 23.8 | 13.1 |

Table 3. Continued

| Cover crop | Early Establishment | Early Plantings[1] Nov. % cover | Early Plantings[1] Cold Tolerance | Early Plantings[1] Nov. height (cm) | Early Plantings[1] Weed Control | Late Plantings[1] Nov. % cover | Late Plantings[1] Cold Tolerance | Late Plantings[1] Nov. height (cm) | Late Plantings[1] Weed Control | COMMENTS |
|---|---|---|---|---|---|---|---|---|---|---|
| Fall rye cv. *Danko* | VF | 85-100 | H | 15-30 | G | 50-80 | H | 5-15 | F | bolts quickly in spring |
| Winter wheat cv. *Monopol* | M-F | 85-100 | H | 20-25 | G | 50-70 | H | 5-15 | F | some powdery mildew |
| Annual ryegrass cv. *Aubade* | VF* | 90-100 | H** | 25-35 | G*** | 60-80 | H | 5-15 | F | very clean of disease |
| Spring wheat cv. *Max* | F | 90-100 | M | 35-55 | G | 45-75 | H | 5-15 | F | planted late; very cold tol. |
| Spring barley cv. *Virden* | VF | 90-100 | L | 50-70 | G | 65-85 | M | 15-30 | F | sig. rust/pow. mildew |
| Winter barley cv. *Elmira* | F | 80-100 | M | 10-15 | G | 65-80 | H | 5-10 | F | sig. rust & brown mould |
| Winter triticale cv. *Wintri* | S-M | 60-80 | H | <5 | F | 20-40 | H | <5 | P | grows close to ground |
| Oats cv. *Cascadia* | VF | 90-100 | L | 45-60 | F | 45-70 | H-M | 15-20 | P | some rust/pow. mildew |
| Forage rape cv. *Liratop* | VF | 90-100 | H | 25-30 | G | 60-90 | H | 5-10 | F | bolts quickly in spring |
| Alsike clover cv. *Common* | S | 40-50 | H | 5-10 | F | N/A# | N/A | N/A | N/A | requires good drainage |
| Red clover cv. *Pacific Red* | S-M | 60-80 | H | 5-10 | F | N/A | N/A | N/A | N/A | requires good drainage |
| Crimson clover cv. *Common* | M | 90-100 | H | 10-15 | F | 20-30 | H | <5 | P | requires good drainage |

[1]See Table 1 for planting dates

*Early establishment: F=fast, M=medium, S=slow, VF=very fast, VS=very slow

**Cold tolerance: H=high, M=moderate, L=low; cover crops with a low cold tolerance usually winter kill; cover crops with a moderate cold tolerance may winter kill; and cover crops with a high cold tolerance do not winter kill and grow will under wet and cool conditions.

***Weed control: P=poor, F=fair, G=good; the most dominant weed was chickweed.

#N/A = not applicable; too slow to establish, therefore not recommended for planting late in growing season.

Table 4. Cover crop screening trials: General field observations for commonly used cover crops.

The true winter cereals (i.e. requiring a cold period for vernalization) were fall rye (Ladner common and the cultivars 'Danko' and 'Kodiac'), winter wheat ('Monopol' and 'Fundelea'), winter barley ('Elmira') and triticale ('Wintri' and 'Pika'). In general, the fall ryes lived up to their reputation for hardiness, disease resistance and growth under cold or cool conditions. By spring plow-down, fall rye ranked among the best cover crops in both dry matter yields and cover. Satisfactory performances were also obtained from early-planted 'Monopol' winter wheat and 'Elmira' winter barley. However, 'Wintri' triticale did not yield well as a green manure crop and did not produce much cover (60-80%) or dry matter and 'Elmira' winter barley and 'Fundelea' winter wheat were heavily infested with fungal diseases.

A successful winter cover crop in this area should establish satisfactory soil cover prior to December and the onset of cold temperatures and high rainfall conditions. Cover, either dead or alive, should be maintained until field operations begin the next spring. It is clear that, aside from the particular crop planted, the date of seeding and soil conditions at seeding are the most important determinants of good cover or biomass yields and residual soil nitrate conservation. Cover crops that established rapidly and provided soil cover, good weed control, low incidence of disease, and relatively good biomass yields over the first two years of this screening trial were: 'Danko' and 'Kodiac' fall rye; 'Monopol' winter wheat; 'Aubade' annual ryegrass; 'Virden' spring barley; and 'Max' spring wheat. Therefore, for the final three years of the screening trials, these cover crops continued to be planted and monitored; and are now the subject of further discussion with respect to their five year comparative performance in green manure production and residual soil nitrate conservation.

### 3.3 Green manure production

Cover crop biomass production prior to winter for the five selected crops is presented in Table 5 for early and late planted cover crops. Delaying planting until the third week of September dramatically reduced fall dry matter production from a range of 2-5 to < 1.5 t ha$^{-1}$. When planted in August, spring barley and spring wheat were the highest yielders, joined by fall rye and annual ryegrass in two of the five years. In most years, early seeded spring wheat and spring barley winter-killed with little additional biomass production after the onset of winter.

For the winter hardy crops, the spring harvest, usually in April, indicates the crops' potential for green manure production (Table 5). For the cover crops seeded during the third week of August, winter wheat consistently produced high amounts of green manure, to a maximum of 10 t ha$^{-1}$ in spring 1995. Annual ryegrass yielded 5 to 9 t ha$^{-1}$ of dry matter prior to spring plow-down. Fall rye usually yielded less biomass than annual ryegrass or winter wheat. Interestingly, spring wheat seeded later consistently survived the winter as did spring barley in two of five years. In fact, spring wheat was the top green manure crop in three of five years and did consistently well in all years.

Annual ryegrass yielded the most green manure of any September seeded cover crop in 1996 and performed consistently well in other years. Winter wheat and fall rye ranked differently from year to year, but were generally lower yielding than spring wheat or annual ryegrass.

Early-planted cover crops produced half of their total biomass prior to winter as compared to late-planted crops, which produced only 5 to 36% of their total biomass prior to winter. Cover crop biomass production prior to winter declined dramatically with later planting dates. Much of the biomass production by the late-planted cereals occurred during late

| | Early planted[1]: approximately the third week of August. | | | | | | | | | | Late planted[1]: approximately the third week of September | | | | | | | | | |
|---|---|---|---|---|---|---|---|---|---|---|---|---|---|---|---|---|---|---|---|---|
| | 1991/92 | | 1992/93 | | 1993/94 | | 1994/95 | | 1995/96 | | 1991/92 | | 1992/93 | | 1993/94 | | 1994/95 | | 1995/96 | |
| Variables | Mean | SD | Mean | SD | Mean | SD | Mean | SD | Mean | SD | Mean | SD | Mean | SD | Mean | SD | Mean | SD | Mean | SD |
| **Fall harvest[1]:** | | | | | | | | | | | | | | | | | | | | |
| Dry weights (t/ha) | p=0.0093 | | p=0.0010 | | p=0.0010 | | p=0.0006 | | p=0.0005 | | p=0.0331 | | p=0.0008 | | p=0.0002 | | p=0.2196 | | | |
| Fall rye | 3.61 | 1.15 * bc* | 1.68 | 0.45 ab | 1.30 | 0.46 c | 2.38 | 0.52 bc | 4.17 | 1.07 a | 0.94 | 0.37 abc | 0.78 | 0.28 a | 0.59 | 0.18 b | 1.57 | 0.72 | NH* | |
| Winter wheat | 2.71 | 0.69 c | 1.07 | 0.57 c | 1.55 | 1.00 c | 1.80 | 0.13 cd | 2.87 | 0.11 b | 0.68 | 0.25 bc | 0.32 | 0.16 b | 0.43 | 0.14 c | NH | | NH | |
| Annual ryegrass | 4.01 | 0.90 bc | 1.43 | 0.31 bc | 2.08 | 0.19 bc | 0.97 | 0.55 d | 4.70 | 1.08 a | 0.57 | 0.19 c | 0.32 | 0.17 b | 0.28 | 0.05 d | NH | | NH | |
| Spring wheat | 5.01 | 0.72 ab | 1.09 | 0.39 c | 3.27 | 1.09 a | 3.11 | 1.08 ab | 4.55 | 0.44 a | 1.14 | 0.30 ab | 0.25 | 0.11 b | 0.50 | 0.08 bc | 1.04 | 0.33 | NH | |
| Spring barley | 5.58 | 1.22 a | 2.08 | 0.88 a | 2.74 | 0.44 ab | 3.68 | 0.41 a | 2.91 | 0.40 b | 1.28 | 0.28 a | 0.57 | 0.11 a | 0.75 | 0.10 a | 1.65 | 0.10 | NH | |
| N concentration (%N) | p=0.0011 | | p=0.0006 | | p=0.0001 | | p=0.0002 | | p=0.3460 | | p=0.6183 | | p=0.0044 | | p=0.0005 | | p=0.0067 | | | |
| Fall rye | 2.60 | 0.44 a | 2.97 | 0.55 cd | 2.25 | 0.21 b | 2.62 | 0.25 bc | 2.61 | 0.32 | 4.61 | 0.66 | 3.88 | 0.49 b | 4.52 | 0.41 b | 3.06 | 0.34 b | NH | |
| Winter wheat | 2.73 | 0.23 a | 3.50 | 0.35 ab | 2.76 | 0.20 a | 3.24 | 0.06 a | 2.70 | 0.57 | 4.56 | 0.34 | 3.16 | 0.78 c | 4.35 | 0.51 b | NH | | NH | |
| Annual ryegrass | 2.74 | 0.13 a | 3.22 | 0.24 bc | 2.20 | 0.12 b | 2.77 | 0.20 b | 2.97 | 0.36 | 4.69 | 0.15 | 4.30 | 0.27 ab | 5.47 | 0.12 a | NH | | NH | |
| Spring wheat | 2.25 | 0.50 a | 3.78 | 0.51 a | 2.13 | 0.30 b | 2.82 | 0.24 b | 2.67 | 0.39 | 4.69 | 0.21 | 4.56 | 0.16 a | 5.23 | 0.31 a | 3.76 | 0.36 a | NH | |
| Spring barley | 1.49 | 0.28 b | 2.65 | 0.23 d | 1.54 | 0.25 c | 2.41 | 0.11 c | 3.12 | 0.27 | 4.31 | 0.30 | 4.02 | 0.12 ab | 4.08 | 0.32 b | 3.72 | 0.10 a | NH | |
| N content (kg/ha) | p=0.0977 | | p=0.0041 | | p=0.0098 | | p=0.0003 | | p=0.0078 | | p=0.0180 | | p=0.0002 | | p=0.0052 | | p=0.1574 | | | |
| Fall rye | 92.9 | 28.0 | 48.9 | 12.8 ab | 29.0 | 10.0 b | 61.3 | 6.6 b | 110 | 35.7 ab | 41.8 | 13.8 ab | 29.3 | 6.2 a | 27.2 | 9.8 a | 47.6 | 21.1 | NH | |
| Winter wheat | 75.0 | 23.1 | 36.8 | 17.0 c | 42.9 | 27.0 b | 58.4 | 5.5 b | 77.4 | 16.2 c | 30.7 | 10.7 b | 10.0 | 2.6 b | 18.5 | 6.3 bc | NH | | NH | |
| Annual ryegrass | 109 | 21.1 | 45.7 | 8.5 ab | 45.7 | 4.6 b | 26.5 | 14.2 c | 138 | 26.8 a | 26.8 | 9.2 b | 13.7 | 2.7 b | 15.3 | 2.3 c | NH | | NH | |
| Spring wheat | 113 | 28.2 | 40.9 | 14.0 bc | 68.8 | 21.5 a | 86.1 | 24.8 a | 123 | 29.3 ab | 53.1 | 11.7 a | 11.5 | 5.2 b | 26.0 | 3.7 ab | 39.7 | 14.8 | NH | |
| Spring barley | 82.1 | 18.9 | 53.7 | 18.5 a | 41.6 | 5.2 b | 88.6 | 9.1 a | 91.4 | 18.8 bc | 54.9 | 10.4 a | 22.9 | 3.6 a | 30.7 | 5.3 a | 61.3 | 4.3 | NH | |

[1]See Table 1 for planting and harvest dates.
#WK = winter kill; NH = insufficient sample to harvest, height less than 5 cm.
**Means in the same column followed by the same letter are not significantly different; alpha=0.05.

Table 5. Selected early and late planted cover crops for fall and spring harvest annual above ground dry weight, nitrogen concentration and content comparisons

| | Early planted[1]: approximately the third week of August. | | | | | | | | | | Late planted[1]: approximately the third week of September | | | | | | | | | |
|---|---|---|---|---|---|---|---|---|---|---|---|---|---|---|---|---|---|---|---|---|
| | 1991/92 | | 1992/93 | | 1993/94 | | 1994/95 | | 1995/96 | | 1991/92 | | 1992/93 | | 1993/94 | | 1994/95 | | 1995/96 | |
| Variables | Mean | SD | Mean | SD | Mean | SD | Mean | SD | Mean | SD | Mean | SD | Mean | SD | Mean | SD | Mean | SD | Mean | SD |
| **Spring harvest[1]:** | | | | | | | | | | | | | | | | | | | | |
| Dry weights (t/ha) | p=0.0133 | | p=0.2902 | | p=0.0359 | | p=0.0015 | | p=0.1450 | | p=0.0027 | | p=0.0112 | | p=0.0005 | | p=0.2873 | | p=0.0001 | |
| Fall rye | 5.19 | 1.05 b | 6.37 | 1.21 | 2.73 | 0.64 b | 5.67 | 0.55 c | 6.46 | 2.33 | 3.98 | 1.06 c | 3.60 | 1.11 ab | 2.81 | 0.68 c | 4.26 | 0.84 | 5.51 | 0.41 b |
| Winter wheat | 7.44 | 2.08 a | 6.80 | 1.64 | 5.79 | 1.25 a | 9.90 | 1.08 a | 9.43 | 3.08 | 6.79 | 1.93 b | 2.17 | 0.55 b | 4.85 | 1.02 b | 5.63 | 0.98 | 4.76 | 1.14 bc |
| Annual ryegrass | 8.89 | 0.61 a | 5.71 | 0.83 | 5.11 | 1.51 a | 8.12 | 1.81 b | 7.40 | 1.77 | 6.84 | 0.90 b | 4.54 | 1.32 a | 4.24 | 0.70 b | 5.18 | 1.97 | 10.8 | 0.64 a |
| Spring wheat | WK* | | 5.27 | 1.13 | WK | | WK | | WK | | 9.11 | 2.08 a | 4.85 | 0.96 a | 6.71 | 0.89 a | 5.54 | 1.08 | 5.45 | 0.32 b |
| Spring barley | WK | | WK | | WK | | WK | | WK | | WK | | WK | | WK | | WK | | 3.92 | 1.37 c |
| N concentration (%N) | p=0.0274 | | p=0.0162 | | p=0.0169 | | p=0.0013 | | p=0.0039 | | p=0.0863 | | p=0.0005 | | p=0.0020 | | p=0.0003 | | p=0.0002 | |
| Fall rye | 1.63 | 0.09 a | 1.07 | 0.12 ab | 1.13 | 0.05 a | 1.30 | 0.15 a | 1.10 | 0.09 a | 1.38 | 0.06 | 1.20 | 0.08 a | 1.12 | 0.03 a | 1.20 | 0.08 a | 1.00 | 0.13 b |
| Winter wheat | 1.43 | 0.13 a | 0.96 | 0.06 b | 1.02 | 0.10 ab | 0.93 | 0.05 b | 0.75 | 0.06 b | 1.15 | 0.08 | 1.13 | 0.05 ab | 0.95 | 0.08 b | 0.89 | 0.14 c | 0.91 | 0.14 bc |
| Annual ryegrass | 1.52 | 0.14 ab | 1.16 | 0.11 a | 0.90 | 0.08 b | 0.98 | 0.03 b | 0.85 | 0.07 b | 1.31 | 0.15 | 1.02 | 0.09 b | 0.90 | 0.01 b | 0.94 | 0.04 bc | 0.80 | 0.02 c |
| Spring wheat | WK | | 1.11 | 0.18 a | WK | | WK | | WK | | 1.21 | 0.13 | 0.86 | 0.07 c | 0.83 | 0.10 b | 1.01 | 0.08 b | 0.91 | 0.14 bc |
| Spring barley | WK | | WK | | WK | | WK | | WK | | WK | | WK | | WK | | WK | | 1.20 | 0.08 a |
| N content (kg/ha) | p=0.0161 | | p=0.4217 | | p=0.0543 | | p=0.1077 | | p=0.6707 | | p=0.0088 | | p=0.0322 | | p=.0215 | | p=0.7297 | | p=0.0001 | |
| Fall rye | 85.4 | 21.4 b | 68.9 | 19.7 | 30.9 | 7.2 | 73.3 | 9.1 | 71.8 | 29.4 | 55.0 | 15.0 c | 43.3 | 13.8 a | 31.5 | 7.9 b | 51.1 | 10.1 | 54.8 | 6.1 b |
| Winter wheat | 107 | 32.9 ab | 65.4 | 17.7 | 58.6 | 11.6 | 91.5 | 10.5 | 69.5 | 21.5 | 77.4 | 19.5 bc | 24.5 | 7.40 b | 46.2 | 11.9 ab | 49.9 | 8.8 | 49.3 | 7.1 b |
| Annual ryegrass | 136 | 20.2 a | 66.1 | 12.0 | 46.8 | 16.7 | 78.9 | 15.7 | 62.1 | 10.4 | 89.5 | 16.9 ab | 45.6 | 11.5 a | 38.1 | 6.0 b | 49.2 | 20.0 | 84.9 | 4.4 a |
| Spring wheat | WK | | 57.2 | 9.3 | WK | | WK | | WK | | 111 | 29.5 a | 42.0 | 9.8 a | 55.3 | 7.8 a | 55.6 | 8.9 | 49.3 | 7.1 b |
| Spring barley | WK | | WK | | WK | | WK | | WK | | WK | | WK | | WK | | WK | | 46.1 | 14.4 b |

Table 5. Continued

March and April. Therefore, the amount of green manure available from late seeded cover crops is strongly dependent on the time available for spring growth prior to establishment of summer crops.

### 3.4 Residual soil nitrate conservation

Cover crop N content was also determined in our screening trials (Table 5). The relative N content rankings closely followed the rankings for biomass production (Table 5). Perhaps the first impression gained from the cover crop N uptake data is the large year to year variability. For example, August seeded annual ryegrass, one of the better N accumulators, took up well over 100 kg N $ha^{-1}$ in two years, but dropped to as little as 26 kg N $ha^{-1}$ in 1994-95. Soil moisture was quite low in August 1994 and the annual ryegrass emergence suffered more from this than the larger seeded cereals. Maximizing cover crop uptake of residual $NO_3$ in the fall requires successful establishment and a long period of good growing conditions prior to December when cold temperatures and low light availability reduce crop growth and N uptake to negligible levels. South coastal BC has more than enough winter rainfall to completely leach any remaining $NO_3$ prior to the onset of the spring growth period (Bomke et al., 1994).

Spring cereals seeded prior to September are usually winter killed and their capability of capturing residual $NO_3$ is minimal when temperatures approach freezing. However, spring barley and wheat can take up substantial residual N, if they have a long enough growth period prior to frost. Also significant is that N concentration at the time of winter-kill usually exceeds 20 g $kg^{-1}$ (2%), the minimum level required for decomposition without immobilizing soil N during the following growing season. The result is that the cover crop N mineralizes when soil temperatures warm in the spring often giving substantially higher available N for crops following winter-killed cereals than over-wintering cover crops (Nafuma, 1998). The latter continue growth under the low soil N conditions that are common following the winter rains.

If cover crop seeding must wait until late September, spring barley is less effective, < 50kg N $ha^{-1}$, in capturing and retaining residual N. Late planted spring wheat is more winter hardy and continues to take up N even following frost if satisfactory growing conditions persist. This occurred during the warm December of the 1991-92 winter. Usually spring wheat, when planted late, behaves like winter cereals, recommencing growth in the spring under low soil N conditions and having crop N concentrations well below the 20 g $kg^{-1}$ minimum required to prevent soil N immobilization. When the over-wintering cover crops were planted in late August there were no significant differences in crop N content except in spring 1992 (Table 5). The potential for growth and N uptake was high during the fall of 1991 and at the spring harvest annual ryegrass contained more N than fall rye but was similar to winter wheat.

Cover crops planted in the third week of September differed significantly in N content the following spring prior to plow-down, except in 1994-95 (Table 5). While differences usually were not large, winter wheat, annual ryegrass and spring wheat were most effective at retaining N over winter, especially in 1991-92, while fall rye was often in the group of crops that contained the least N content by spring plow down time. For both the early and late plantings, such results are also reflected in the fall versus spring harvest ratios and differences with respect to biomass, N concentration and content (Table 6). For the rapidly establishing and relatively more cold tolerant fall rye, its fall/spring harvest biomass and N content ratios were greater than the other over-wintering cover crops except in 1995-96.

| | Early planted[1]: approximately the third week of August. | | | | | | | | | | Late planted[1]: approximately the third week of September | | | | | | | |
|---|---|---|---|---|---|---|---|---|---|---|---|---|---|---|---|---|---|---|
| Year*: | 1991/92 | | 1992/93 | | 1993/94 | | 1994/95 | | 1995/96 | | 1991/92 | | 1992/93 | | 1993/94 | | 1994/95 | |
| Variables | Mean | SD | Mean | SD | Mean | SD | Mean | SD | Mean | SD | Mean | SD | Mean | SD | Mean | SD | Mean | SD |
| **Fall vs. spring harvest ratios[1]:** | | | | | | | | | | | | | | | | | | |
| Dry weights | p=0.0380 | | p=0.0290 | | p=0.0700 | | p=0.0039 | | p=0.1172 | | p=0.1055 | | p=0.0359 | | p=0.0034 | | p=0.1459 | |
| Fall rye | 0.69 | 0.16 a** | 0.26 | 0.03 a | 0.50 | 0.21 | 0.43 | 0.13 a | 0.72 | 0.30 | 0.25 | 0.14 | 0.24 | 0.14 a | 0.21 | 0.08 a | 0.39 | 0.21 |
| Winter wheat | 0.38 | 0.12 b | 0.15 | 0.05 b | 0.26 | 0.15 | 0.18 | 0.01 b | 0.33 | 0.12 | 0.11 | 0.07 | 0.15 | 0.03 ab | 0.09 | 0.05 b | n/a | |
| Annual ryegrass | 0.46 | 0.13 b | 0.25 | 0.05 a | 0.44 | 0.14 | 0.13 | 0.09 b | 0.67 | 0.26 | 0.08 | 0.03 | 0.08 | 0.04 b | 0.07 | 0.02 b | n/a | |
| Spring wheat | n/a # | | 0.22 | 0.11 ab | n/a | | n/a | | n/a | | 0.13 | 0.02 | 0.06 | 0.03 b | 0.08 | 0.02 b | 0.19 | 0.04 |
| Spring barley | n/a | | n/a | | n/a | | n/a | | n/a | | n/a | | n/a | | n/a | | n/a | |
| Nitrogen concentration | p=0.2389 | | p=0.0036 | | p=0.0234 | | p=0.0001 | | p=0.1110 | | p=0.0782 | | p=0.0002 | | p=0.0015 | | p=0.0140 | |
| Fall rye | 1.60 | 0.28 | 2.82 | 0.71 b | 1.99 | 0.20 b | 2.03 | 0.23 c | 2.38 | 0.25 | 3.34 | 0.41 | 3.24 | 0.47 c | 4.04 | 0.29 b | 2.56 | 0.30 b |
| Winter wheat | 1.92 | 0.27 | 3.67 | 0.41 a | 2.73 | 0.34 a | 3.52 | 0.26 a | 3.66 | 0.86 | 3.96 | 0.35 | 2.79 | 0.61 c | 4.63 | 0.66 b | n/a | |
| Annual ryegrass | 1.82 | 0.20 | 2.81 | 0.44 b | 2.44 | 0.11 ab | 2.85 | 0.27 b | 3.54 | 0.71 | 3.62 | 0.35 | 4.23 | 0.21 b | 6.08 | 0.18 a | n/a | |
| Spring wheat | n/a | | 3.48 | 0.74 a | n/a | | n/a | | n/a | | 3.90 | 0.35 | 5.31 | 0.37 a | 6.41 | 0.98 a | 3.72 | 0.36 a |
| Spring barley | n/a | | n/a | | n/a | | n/a | | n/a | | n/a | | n/a | | n/a | | n/a | |
| Nitrogen content | p=0.1632 | | p=0.2141 | | p=0.2340 | | p=0.0297 | | p=0.2088 | | p=0.1159 | | p=0.0393 | | p=0.0440 | | p=0.3368 | |
| Fall rye | 1.09 | 0.22 | 0.72 | 0.12 | 0.97 | 0.39 | 0.86 | 0.22 a | 1.76 | 0.94 | 0.80 | 0.36 | 0.74 | 0.32 a | 0.87 | 0.33 a | 0.96 | 0.46 |
| Winter wheat | 0.73 | 0.26 | 0.54 | 0.13 | 0.73 | 0.43 | 0.64 | 0.09 ab | 1.25 | 0.59 | 0.43 | 0.25 | 0.41 | 0.10 b | 0.44 | 0.24 b | n/a | |
| Annual ryegrass | 0.84 | 0.29 | 0.70 | 0.11 | 1.08 | 0.37 | 0.36 | 0.22 b | 2.28 | 0.65 | 0.30 | 0.05 | 0.32 | 0.15 b | 0.41 | 0.11 b | n/a | |
| Spring wheat | n/a | | 0.72 | 0.24 | n/a | | n/a | | n/a | | 0.50 | 0.14 | 0.29 | 0.16 b | 0.48 | 0.09 b | 0.70 | 0.18 |
| Spring barley | n/a | | n/a | | n/a | | n/a | | n/a | | n/a | | n/a | | n/a | | n/a | |

[1]See Table 1 for planting and harvest dates.

#n/a: spring harvest cover crop either winter kill or there was insufficient sample to harvest in the fall (height less than 5 cm); see Table 5.

*1995/96 late planted cover crops were not harvested in the fall; see Table 5.

**Means in the same column followed by the same letter are not significantly different; alpha=0.05.

Table 6. Selected early and late planted cover crops fall vs. spring harvests ratios and differences for annual above ground dry weight, nitrogen concentration and content.

| Year*: | Early planted[1]: approximately the third week of August. | | | | | | | | | | Late planted[1]: approximately the third week of September | | | | | | | |
|---|---|---|---|---|---|---|---|---|---|---|---|---|---|---|---|---|---|---|
| | 1991/92 | | 1992/93 | | 1993/94 | | 1994/95 | | 1995/96 | | 1991/92 | | 1992/93 | | 1993/94 | | 1994/95 | |
| Variables | Mean | SD | Mean | SD | Mean | SD | Mean | SD | Mean | SD | Mean | SD | Mean | SD | Mean | SD | Mean | SD |
| **Fall vs. spring harvest differences[1]:** | | | | | | | | | | | | | | | | | | |
| Dry weights | p=0.0127 | | p=0.0829 | | p=0.0342 | | p=0.0016 | | p=0.0850 | | p=0.0038 | | p=0.0134 | | p=0.0007 | | p=0.0325 | |
| Fall rye | 1.58 | 0.81 b | 4.69 | 0.77 | 1.43 | 0.77 b | 3.29 | 1.01 b | 2.29 | 2.06 | 3.04 | 1.06 b | 2.82 | 1.28 bc | 2.22 | 0.61 c | 2.69 | 1.38 b |
| Winter wheat | 4.73 | 1.82 a | 5.73 | 1.07 | 4.24 | 1.16 a | 8.10 | 0.99 a | 6.56 | 3.13 | 6.11 | 2.02 a | 1.85 | 0.09 c | 4.42 | 1.06 b | n/a | |
| Annual ryegrass | 4.88 | 1.41 a | 4.28 | 0.79 | 3.03 | 1.69 ab | 7.15 | 2.21 a | 2.70 | 2.33 | 6.27 | 0.83 a | 4.22 | 1.34 ab | 3.96 | 0.73 b | n/a | |
| Spring wheat | n/a | | n/a | | n/a | | n/a | | n/a | | 7.97 | 1.86 a | 4.60 | 0.96 a | 6.21 | 0.93 a | 4.50 | 0.87 a |
| Spring barley | n/a | | n/a | | n/a | | n/a | | n/a | | n/a | | n/a | | n/a | | n/a | |
| Nitrogen concentration | p=0.3790 | | p=0.0243 | | p=0.0110 | | p=0.0001 | | p=0.2949 | | p=0.7615 | | p=0.0030 | | p=0.0032 | | p=0.0197 | |
| Fall rye | 0.97 | 0.46 | 1.90 | 0.58 b | 1.12 | 0.21 b | 1.32 | 0.24 c | 1.51 | 0.27 | 3.23 | 0.62 | 2.68 | 0.49 bc | 3.40 | 0.39 b | 1.87 | 0.34 b |
| Winter wheat | 1.30 | 0.29 | 2.54 | 0.34 a | 1.74 | 0.25 a | 2.32 | 0.11 a | 1.96 | 0.58 | 3.40 | 0.33 | 2.03 | 0.75 c | 3.40 | 0.53 b | n/a | |
| Annual ryegrass | 1.22 | 0.20 | 2.06 | 0.33 b | 1.30 | 0.04 b | 1.80 | 0.22 b | 2.12 | 0.42 | 3.38 | 0.17 | 3.28 | 0.20 ab | 4.57 | 0.12 a | n/a | |
| Spring wheat | n/a | | n/a | | n/a | | n/a | | n/a | | 3.48 | 0.17 | 3.70 | 0.13 a | 4.40 | 0.39 a | 2.75 | 0.34 a |
| Spring barley | n/a | | n/a | | n/a | | n/a | | n/a | | n/a | | n/a | | n/a | | n/a | |
| Nitrogen content | p=0.1173 | | p=0.2913 | | p=0.2513 | | p=0.0550 | | p=0.0791 | | p=0.0268 | | p=0.0808 | | p=0.0397 | | p=0.3802 | |
| Fall rye | -7.53 | 15.3 | 20.0 | 10.2 | 1.89 | 10.6 | 12.0 | 15.5 | -38.5 | 33.3 | 13.2 | 17.1 b | 14.0 | 18.2 | 4.30 | 8.8 b | 3.5 | 24.9 |
| Winter wheat | 31.8 | 30.5 | 28.7 | 4.6 | 15.7 | 25.9 | 33.2 | 10.3 | -7.90 | 37.3 | 46.7 | 25.6 a | 14.5 | 24.9 | 27.7 | 15.0 a | n/a | |
| Annual ryegrass | 26.5 | 37.6 | 20.4 | 9.2 | 1.08 | 18.9 | 52.3 | 26.1 | -76.2 | 28.2 | 62.7 | 10.4 a | 32.0 | 13.0 | 22.8 | 7.8 a | n/a | |
| Spring wheat | n/a | | n/a | | n/a | | n/a | | n/a | | 57.8 | 25.7 a | 30.6 | 10.6 | 29.3 | 7.5 a | 16.0 | 8.3 |
| Spring barley | n/a | | n/a | | n/a | | n/a | | n/a | | n/a | | n/a | | n/a | | n/a | |

Table 6. Continued

Furthermore, as discussed above, the fall rye biomass and N content differences were always amongst the least relative to the other mentioned over-wintering cover crops, particularly for the late plantings. Therefore, both harvest ratios and differences between spring and fall values suggest that for fall rye a greater proportion of its cumulative biomass and N content is accumulated in the fall, whereas a greater proportion of the biomass and N contents for the other over-wintering cover crops resulted during spring growth. This was particularly evident with the late planted cover crops' N content ratios and differences, where the fall rye N content ratios were highest and differences much less. These results suggest that while the fall rye was particularly effective in N uptake during the fall, its ability to retain such N is less than other cover crops, such as winter wheat, annual ryegrass and spring wheat. Such results could be related to the fall rye losing a relatively greater amount of biomass over the winter via the loss of older shoots or tillers.

Early planted cover crops usually captured more residual N than when planted late. Nitrogen uptake often exceeded 75 kg $ha^{-1}$ for cover crops planted in August as compared to an average closer to 50 kg $ha^{-1}$ for most of the crops when planted in the third week of September. Cover crops cannot be relied upon to prevent $NO_3$ leaching if N additions to the previous crop are greatly in excess of crop needs or if the cover crop growth potential prior to winter is low. With appropriate fertilizer rates applied to the previous cash crop, the cover crops can reduce soil residual $NO_3$ to low levels and, over the longer term prevent the loss of substantial quantities of soil N. Preventing leaching losses of 500 to 750 kg N $ha^{-1}$ during a decade of cover cropping will result in substantially higher mineralizable soil N and reduced fertilizer requirements. The soils upon which our experiments were conducted contain between 2000 and 4000 kg $ha^{-1}$ of total N, most of which is stable and does not enter the pool of available N. Forms of N that are readily mineralized into available N for summer crops are dominated by fresh crop residue N such as that which is retained by a sustained cover cropping program.

The problem of short-term immobilization of available soil N by a decomposing cover crop is of concern. Any non-leguminous cover crop that is allowed to grow during the spring in south coastal BC will have low N concentrations by plow-down and will reduce short term N availability to the subsequent crop. The farmer may compensate for this by:

- killing the cover crop early (and losing some of the green manure value of the cover crop);
- increasing external N inputs (e.g. fertilizers or manures); or
- adding a legume to the over-winter cover crop and using biological N fixation during the spring growth period to increase the concentration of N in the cover crop (Odhiambo & Bomke, 2000).

## 4. Conclusions

Relative to the cover crops screened, fall rye, annual ryegrass, winter wheat, spring wheat and spring barley performed the best. Planting cover crops early (late August) consistently gave two to four times higher biomass yields prior to winter than the late-planted (late September) cover crops and spring cereals often produced greater biomass than other cover crops. Early seeded spring cereals usually winter-killed and produced between 3 and 5 t $ha^{-1}$ of dead mulch material; while early-seeded, over-wintering cover crops produced 5 to 8 t $ha^{-1}$ of green manure in the spring. Late-seeded, over-wintering cover crops produced 4 to 6 t $ha^{-1}$ of green manure with 'Max' spring wheat over-wintering and performing consistently well. With respect to residual soil nitrate conservation, comparisons of relative N contents (rankings) closely followed those for biomass yields. Delaying cover crop seeding until late September reduced the uptake of residual soil N. All late seeded cover crops took up less than 60 kg $ha^{-1}$ of N prior to winter, but the performance of each cover crop varied greatly among years.

## 5. Acknowledgments

This project would have not been possible without the co-operation of the many farmers in Delta, specifically Bert Nottingham, Rod Swenson, Dennis Kamlah, Ab Singh, Gordon Ellis and Hugh and Stan Reynolds. Funding was provided by the Canada - British Columbia Soil Conservation Agreement and The Delta Farmland & Wildlife Trust through the Canada - British Columbia Green Plan for Agriculture, cost shared equally by the governments of Canada and British Columbia and administered by Agriculture Canada and the British Columbia Ministry of Agriculture, Fisheries and Food. Funding for the cost-shared cover crop program, the "Greenfields" project, was primarily provided by participating farmers, Canadian Wildlife Service (CWS) of Environment Canada and Ducks Unlimited Canada. The authors also wish to acknowledge the help of Rick McKelvey (CWS), Sean Boyd (CWS), Sabine Neels (DF&WT), Adriele Park (University of British Columbia) and Katrina Nolan (University of British Columbia) for their comments and editorials with respect to the publication of this document.

## 6. References

Bomke, A.A.; Yu, S. & Temple, W.D. (1994). Winter wheat growth and nitrogen demand in south coastal British Columbia. *Canadian Journal of Soil Science* 74(4): 443-451.

Hermawan, B. (1995). Soil structure associated with cover crops and grass leys in degraded lowland soils of Delta. Ph.D. Thesis, University of British Columbia, Vancouver, B.C. 154 pp.

Hermawan, B. & Bomke, A.A. (1996). Aggregation of a degraded lowland soil during restoration with different cropping and drainage regimes. *Soil Technology* 9:239-250.

Klohn, L.; Holm and Associates. & Runka, C.G. (1992). Delta Agricultural Study. Agri-Food Regional Development Subsidiary Agreement, Vancouver, B.C. 130pp.

Liu, A. (1995). Soil organic components and aggregation as influenced by cover and ley crops. M.Sc. Thesis, University of British Columbia, Vancouver, B.C. 77pp.

Luttmerding, H.A. (1981). Soils of Langley-Vancouver map area. RAB Bulletin 18, Vol.3 B.C. Ministry of Environment Assessment and Planning Div., Kelowna, B.C. 227pp.

Nafuma, L. (1998). Short-term effects of graminaceous cover crops on autumn soil mineral nitrogen cycling on western lower Fraser valley soils. Ph.D. Thesis. University of British Columbia, Vancouver, B.C. 185 pp.

Odhiambo, J.J.O. & Bomke, A.A. (2000). Short term nitrogen availability following overwinter cereal/grass and legume cover crop monoculture and mixtures in South Coastal British Columbia. *Journal of Soil and Water Conservation.* 55 (3): 347-354.

Parkinson, J.A. & Allen, S.E. (1975). A wet digestion procedure suitable for the determination of nitrogen and nutrients in biological material. *Communications in Soil Science and Plant Analysis* 6:1-11.

Technicon. (1974). Technicon Autoanalyzer II. Methods of nitrogen and phosphorus in BD digest. Industrial method No. 334-74A. Technicon Systems, Terrytown, NY.

Temple, W.D.; Bomke, A.A. & Duynstee, T. (1991). Winter cover crop management in a high rainfall region with large populations of waterfowl. In: *Proceedings of Cover Crops for Clean Water Conference*. April 9-11, 1991. Jackson TN, USA. pp.196-198.

# 4

# Soil Management Strategies for Radish and Potato Crops: Yield Response and Economical Productivity in the Relation to Organic Fertilizer and Ridging Practice

Masakazu Komatsuzaki[1] and Lei Dou[1,2]
[1]*College of Agriculture, Ibaraki University, Ibaraki*
[2]*Soil and Fertilizer Institute, Shandong*
[1]*Japan*
[2]*China*

## 1. Introduction

Modern conventional agriculture, which is based on heavy use of chemical fertilizers and energy inputs and oriented to maximization of returns and profits, results in soil fertility decrease, ground water pollution, ecological environment unbalance, as often regarded as detrimental and unsustainable when considered from economic, social and environmental perspectives in a long term. Since the beginning of the 21st century, as sustainability awareness spreads, organic farming and organic food have been attracting a lot of attention due to their direct relationships to nutrition and environment, both essential for our future life. In this context, the development of organic farming has showed a strong growth in the world. Organic farming is the process that crops are planted using only natural methods to increase yield and maintain soil fertility. In comparison with conventional agriculture, organic farming could produce healthier agricultural products while conserving the quality of the soil and surrounding environment.

The transition from conventional to organic farming is accompanied by changes in an array of soil chemical and physical properties and processes that affect soil fertility (Jongtae Lee, 2010). Organic management forms the foundation of an healthy and sustainable agricultural system. Organic farming regenerates soil fertility through organic fertilizer and proper farming practices, which could increase the organic matter in the soil. Increased organic matter makes nutrients more available, improves soil structure, raises crop growth, enhances water field capacity and drainage, and decreases soil erosion(Brian Baker et al, 2005).

Organic fertilizer has been alternatively attention since it enhances soil quality, especially adding soil organic matter and ensures activities of soil micro livings. Crops require more than 20 nutrients for growth, and some of these nutrients are obtained primarily from the soil. Organic fertilizer contains significantly high levels of nutrients, and is also high in organic matter content and a variety of micronutrients in general, which may improve the crop yield in the long term. At the same time, the utilization of organic fertilizer helps to reduce resource use, because it is usually made of organic by-products, such as oil lees, soybean meal, and fish meal. Organic fertilizer mainly relies on renewable materials rather than on nonrenewable materials and fossil fuels. Organic fertilizer is thus an important material for organic, chemical free agriculture and eco-farming.

The demand of organic agricultural products is increasing rapidly since modern agriculture has been said to be degrading environmental quality and food safety. In 1999, the JAS law defined the organic agricultural products that must be certified by an organic certification group. This has led to a greater demand for organic products. Today, the amount of current organic agricultural products lags behind consumer demand. However, the amount of chemical free agricultural products has been increasing because they are easier to grow than organic products. The demand is especially high for processed foods, such as potato chips, yellow radish pickled, and so on.

One of the major problems impeding the use of organic fertilizer is that it is usually more expensive than chemical fertilizer. However, if organic fertilizer can increase the crop yields over the long term by enhancing soil quality, rising costs for using organic fertilizer may be offset by increasing yields. However, the yield produced by organic fertilizer would differ depending on the crop, soil, and climatic conditions.

Potatoes and radish require a high level of soil fertility and these are important crops in the rotation in Japan. Recently, traditional composting practice that called *Bocashi* are becoming widely spread in Asian countries. One of the major advantages of using *Bocashi* is enhancing the recycle of agricultural by-products, and reduces the costs for keeping nutrient in the soil. However, the information regarding yield response and qualities of potato using *Bocashi* are limited. For potato production, ridging practice affects the yield response and economical productivities (Carter et al. 2005), and ridging practice may affects the decompose speed of organic fertilizer due to the difference of soil temperature in the ridges.

The objectives of this research were to determine the yield response of using organic fertilizer, manure, and effective microbial materials for radish and potato cultures. These results would be useful for making cost analyses between organic and chemical fertilizer application.

## 2. Materials and methods

### 2.1 Yield response between chemical and organic fertilizer for radish and potato culture from 1994 through 1997

#### 2.1.1 Field and experimental design

The research of radish was conducted at the experimental farm of Ibaraki University (latitude 36 ° 1'48"N, longitude 140°12'40"E) in Humic Allophane soils (Hapluduat, Haplic Andosols) from 1994 through 1997. The three fertilizer application systems (chemical, organic, and organic with microbial materials), with cow manure application represented the split factors. The fertilizer and manure application systems were shown in Table 1.

The organic fertilizer contained soybean and rapeseed meal, fish meal, bone and brood meal (Source from Iseki Kanto, Co. Ltd.). Microbial materials is the meat meal, rice bran, wheat bran, and microorganism included bacteria (*Azotomonas, Bacillus, Pseudomonas, Clostridium,*

*Rhizobium* and *Bacterium*), yeast (*Saccharomyces, Candida, Endomycopis,* and *Cladosporium*), actinomyces (*Nocardia* and *Streptomyces*) , and a filamentous fungus (*Aspergillus*) (Source from Iseki Kanto, Co. Ltd.). Total nitrogen and carbon of these organic materials were shown in table 2.

| Fertilizer | Microbial material | Manure | Input amount (kg ha$^{-1}$) | | | | Contents (kg ha$^{-1}$) | | |
|---|---|---|---|---|---|---|---|---|---|
| | | | Manure | Chemical fertilizer[1)] | Organic fertilizer[2)] | Microbial material[3)] | N | $P_2O_5$ | $K_2O$ |
| Chemical | - | - | - | 2000 | - | - | 160 | 300 | 140 |
| Chemical | - | + | 10,000 | 2000 | - | - | 160 | 300 | 140 |
| Organic | - | - | - | - | 2000 | - | 160 | 300 | 140 |
| Organic | - | + | 10,000 | - | 2000 | - | 160 | 300 | 140 |
| Organic | + | - | - | - | 2000 | 600 | 160 | 300 | 140 |
| Organic | + | + | 10,000 | - | 2000 | 600 | 160 | 300 | 140 |

Note: 1) Chemical fertilizer was blended at 8, 15, and 7 % of N, $P_2O_5$, and $K_2O$, respectively.
2) Organic fertilizer was also blended at the same contents as chemical fertilizer.

Table 1. Fertilizer materials and input amount at each plot

| Material | Total N (%) | Total C (%) | C/N ratio |
|---|---|---|---|
| Organic fertilizer | 8.0 | 37.5 | 4.7 |
| Microbial material | 2.2 | 14.2 | 65.0 |
| Manure | 1.7 | 21.0 | 12.4 |

Table 2. Total nitrogen and carbon contents of organic materials

Each plot was 3m wide and 6m long with 2 replications. The field was harvested soybean in 1993, and tilled by rotary before planting radishes in 1994. Each plot was tilling by deep tiled rotary after appropriate fertilizer and manure application at each crop planting date.

### 2.1.2 Radish and potato culture

Double crop radish (*Raphanus sativus* L). was planted in the spring (summer radish) and in autumn (autumn radish) of each year. Summer radishes(cv. Taibyou-souhutori) were planted with a plastic mulch on 26, April,1995, 24, April 1996, 18, April 1997. Seeds were planted in a 25cm × 25cm zigzag pattern. The crops were harvested on 22 June ,1995; 25 June 1996 and 18 June 1997. Autumn radishes(cv. Taibyou-souhutori) were planted on 12 September, 1994; 10 September, 1995; 9 September,1996; 10 September,1997. Planting interval was 60cm × 20cm. These were harvested on 12 December, 1994; 15 December,1995; 13 December,1996 and 5 December,1997. For each radish culture, insecticide (DDVP) was sprayed for each growing season to prevent cabbage armyworm (*Manestra brassicae L.*) and common cabbage worm (*Pieris rapae crucivora Boisduval*).

The plots were divided into two sites after radish harvest in 1996. Potato (cv. Tyoshiro) was planted on 25 March, 1997 with a 72cm × 25cm interval after the same fertilizer treatment had been applied. Ridging treatment was applied on 30 April, and harvested on 24 June. No herbicide or insecticide was applied for potato culture.

### 2.1.3 Measurement

Radish yields were measured at each harvest time. A continuous of 20 radishes in a row was taken as a sample at each plot. Each radish was measured for fresh root weight, fresh leaf weight, and root length. At the same time, irregular roots were marked as unmarketable products.

Potato plant length was measured at 30 days after planting and flowering in 1997. Above and potato biomass were determined by sampling 20 continuous potato plants, as potato yields. Potatoes that were over 30g were measured as marketable products.

Soil nitrate concentration was measured on 30 Apr, 23 May, and 24 June in 1997during the potato cropping period. Soils were air dried, and extracted with distilled water. Soil nitrate concentrations were measured by nitrate ion analyzer.

Soil samples for determining microbial populations were taken at 10cm depth in each plot after the summer radish harvest on 5 Jul ,1997 to determine the populations of soil bacteria, fungus, and actinomyces. Microbial population size was determined using the dilution agar-plate method. Rose bengal agar media was used for filamentous fungus test. For the bacteria and actinomycetes test, albumin agar media were used for tests.

## 2.2 Yield response between different ridging managements for organic fertilizer based potato culture in 2010

### 2.2.1 Field and experimental design

A field experiment of potato production was conducted at the farmer's farm (latitude 35°56'35"N, longitude140°10'36"E) in Ushiku City, Ibaraki prefecture, Japan in clay loam humic allophane soils. *Bokashi* and chemical fertilizers were applied in potato production. Bokashi was developed in Japan and it uses microorganisms to ferment the waste into nutrients rich compost, was widely applied to increase soil organic matter and improve microbe's activity. *Bokashi* was made from rice bran, rapeseed meal, rice husk, EM1, sugar and water. The nutrients content of *bokashi* fertilizer were shown in Table 3.

| Nutrients | Ca (mg/100g) | K (mg/100g) | Mg (mg/100g) | $NH_4^+$-N (mg/100g) | $NO_3^-$-N (mg/100g) | P% | C% | N% | C/N |
|---|---|---|---|---|---|---|---|---|---|
| *Bokashi* | 2 | 52 | 7 | 0.8 | 115.6 | 34.6% | 45.22 | 4.59 | 9.85 |

Table 3. Nutrients content of *Bokashi* fertilizer

Potato was planted in a 55m × 53m area, and planting interval was 0.78m ×55m. Six 55m×0.78m plots were laid out, consisting of two treatments with three replications. One treatment was ridged two times+ Inter-tillage (Ridging 3) and another treatment was ridged two times (Ridging 2).The shapes of ridging were measured by leica DISTO™ A6, and the data were recorded every 1 cm. The ridging management was different resulting in the different ridging shapes.

### 2.2.2 Potato culture

Potato was planted in April and harvested in July, 2010. The cropping schedule and farming practice schedule were shown in Table 4. The total rainfall and mean temperature during potato cropping were 511.5 mm and 19.3°C, respectively (Japan Meteorological Agency, 2011).

| Dates | Farm Work and Experimental Contents | Materials and Machine | Ridging 3 | Ridging 2 |
|---|---|---|---|---|
| 2010/2/10 | Fertilizer application | *Bokashi* 1000kg ha$^{-1}$ | + | + |
| | Subsoiling | Tractor (60PS) + Subsoiler (W 130cm) | + | + |
| | plow | Tractor (60PS) + Plow(depth 25cm) | + | + |
| | harrow | Tractor (60PS) + Harrow(w 230cm) | + | + |
| | harrow | Tractor (28PS) + Rotary harrow(w 220cm) | + | + |
| 2010/2/20 | Seed preparation | Seed potato (mother tuber), 1400kg | + | + |
| | Fungicide | | + | + |
| 2010/3/5 | Planting | Seed potato (mother tuber), 1400kg Planter | + | + |
| 2010/3/15 | Inter-tillage | Walking cultivator | + | - |
| 2010/3/25 | Weeding | Weeder (Working type) | + | + |
| | Fertilizer application | NPK (10-16-14) 200kg Hand spray | + | + |
| 2010/4/10 | Inter-tillage | Walking cultivator (upper type) | + | – |
| 2010/5/5 | Pesticide spray | Sprayer | + | + |
| 2010/5/10 | Ridging | Walking cultivator + Ridger | + | + |
| | Fertilizer application | PK (0-20-15) 200kg Hand spray | + | + |
| 2010/5/20 | Pesticide spray | Sprayer | + | + |
| | Ridging | Walking cultivator + Ridger | + | + |
| 2010/6/5 | Pesticide spray | Sprayer | + | + |
| 2010/6/20 | Pesticide spray | Sprayer | + | + |
| 2010/7/10 | Harvest | Potato harvester | + | + |

"+" meant applied, "-" meant did not apply

Table 4. Cropping schedule and farming system in potato production.

### 2.2.3 Measurement

Potato tuber, root and shoot of potato plant were taken from each experimental plots in flowering season on June 2nd, 2010. Ten continuous potato plant's height was measured in each experimental plot. The fresh weight of potato tuber was measured by electrical balance, the shoot and root were oven dry at 60°C for 72 hours and weighted to record the biomass. Matured potatoes were harvested on July 8, 2010 to calculate tuber yield and number. To avoid possible edge effects, in each experimental plot the inner ten potato plants were harvested manually as sample plants. Each potato was weighted to calculate starch value (%) and graded sizes based on the weight to assess appearance quality relevant for marketable value.

Soil samples were taken from topsoil (0-0.10m) on: 1) April 13th,2010 after potato planting; 2) April 26, May 17 at vegetative growth stage; 3) June 2, June 18 in flowering season; 4) July 2 and July 8 at tuber maturation stage. Soil samples were collected at each experimental plot. The collected soil samples were air dried, pulverized, ground and sieved to pass 2 mm mesh before being sent to laboratory for soil nutrients concentration analysis. Soil nutrients analysis of the samples was conducted by SPCA analyzer in FS Center, College of Agriculture, Ibaraki University, Japan.

## 3. Results & discussion

### 3.1 Radish yield responses and related analysis from 1994 to 1997

Average radish root weights from 1$^{st}$ to 7$^{th}$ harvest is shown in Table 5. At 1st harvest, there were significant differences between fertilizer type and manure input for leaf weight, organic fertilizer and manure plots showing higher leaf weight than the chemical fertilizer and no manure input plots at 1st harvest. However, there were no significant differences in root weight. At 2nd and 3rd harvests, there were no significant differences between fertilizer system and manure treatment. Following 4th harvest, root growth weight in the organic fertilizer plots was significantly higher than chemical fertilizer plots. For example, across the manure treatments, average roots weight were 964g for organic fertilizer, 1010g for organic plus microbial material, and 789g for chemical fertilizer at 7th harvest.

There were no significant differences between chemical and organic fertilizer for 1 or 2 years, however, it was clear that the radish growth in the organic fertilizer plots was much higher compared with chemical fertilizer plots 2 or 3 years later. For manure treatment, manure input enhanced leaf growth, but did not show significant effects for root growth.

Noguchi (1992a) reported that organic fertilizer released inorganic nitrogen slower than chemical fertilizer, but it faster than manure. For example, 60-40% organic nitrogen of organic fertilizer released inorganic nitrogen to 200 days at 30°C (Nouguchi, 1992b). So organic fertilizer application over the long term would help to increase soil nutrients. In this research, changes in soil organic matter and nitrogen content were unknown, however, significant crop yield responses may show the difference of soil nutrient contents between organically and chemical fertilizer system.

Average data for 3 years are shown in Table 5. For summer radish culture, the leaf weight was higher in manure plots than no manure input plots, but there was little difference in root length and weight among fertilizer application systems. For autumn radish, the leaf and root weights were higher in organic fertilizer and microbial material plots than in the chemical fertilizer plots. In most cases, root weight decrease in manure plots in each fertilizer application system. However, the effect of microbial materials on radish growth was not clear.

| Fertilizer | Micro-bial material | Manure | 1st (autumn) Leaf weight (g/plant) | 1st (autumn) Root weight (g/plant) | 2nd (summer) Leaf weight (g/plant) | 2nd (summer) Root weight (g/plant) | 3rd (autumn) Leaf weight (g/plant) | 3rd (autumn) Root weight (g/plant) | 4th (summer) Leaf weight (g/plant) | 4th (summer) Root weight (g/plant) |
|---|---|---|---|---|---|---|---|---|---|---|
| Chemical | - | - | 333.3 | 1155.4 | 578.5 | 912.0 | 378.8 | 876.3 | 384.4 | 767.5 |
| Chemical | - | + | 310.6 | 1046.2 | 620.0 | 939.5 | 399.5 | 949.3 | 425.5 | 716.0 |
| Organic | - | - | 312.6 | 1077.0 | 597.3 | 973.5 | 380.0 | 935.5 | 425.0 | 774.9 |
| Organic | - | + | 403.2 | 1198.7 | 603.3 | 817.5 | 356.8 | 871.0 | 429.0 | 756.9 |
| Organic | + | - | 372.4 | 1231.2 | 597.8 | 836.0 | 393.0 | 1070.3 | 368.8 | 847.4 |
| Organic | + | + | 425.8 | 1118.2 | 579.5 | 784.0 | 363.0 | 880.5 | 493.9 | 886.9 |
| Significa nt | | | | | | | | | | |
| Fertilizer type | | | * | N.S. | N.S. | N.S. | N.S. | N.S. | N.S. | * |
| Manure | | | * | N.S. | N.S. | N.S. | N.S. | N.S. | * | N.S. |

| Fertilizer | Micro-bial material | Manure | 5th (autumn) Leaf weight (g/plant) | 5th (autumn) Root weight (g/plant) | 6th (summer) Leaf weight (g/plant) | 6th (summer) Root weight (g/plant) | 7th (autumn) Leaf weight (g/plant) | 7th (autumn) Root weight (g/plant) |
|---|---|---|---|---|---|---|---|---|
| Chemical | - | - | 295.0 | 683.2 | 590.2 | 877.9 | 382.3 | 1610.9 |
| Chemical | - | + | 294.2 | 648.4 | 607.3 | 701.7 | 408.2 | 1509.3 |
| Organic | - | - | 307.8 | 805.7 | 661.6 | 956.8 | 414.6 | 1677.9 |
| Organic | - | + | 340.8 | 752.5 | 695.0 | 972.0 | 410.4 | 1537.8 |
| Organic | + | - | 342.7 | 776.6 | 580.4 | 970.0 | 417.4 | 1669.6 |
| Organic | + | + | 332.0 | 685.6 | 726.2 | 1050.9 | 411.2 | 1483.5 |
| Significant | | | | | | | | |
| Fertilizer type | | | * | * | * | * | N.S. | N.S. |
| Manure | | | N.S. | N.S. | * | N.S. | N.S. | * |

Table 5. The average of radish growth in the relation to fertilizer system (3years)

| Fertilizer | Microbial material | Manure | Soil microorganism Bacteria ($\times 10^7$/g) | ($\times 10^7$/g) | Fungus ($\times 10^5$/g) | B/F ratio |
|---|---|---|---|---|---|---|
| Chemical | - | - | 2.2 | 2.8 | 4.1 | 53.7 |
| Chemical | - | + | 3.9 | 2.1 | 2.8 | 139.3 |
| Organic | - | - | 2.4 | 3.3 | 2.9 | 82.8 |
| Organic | - | + | 4.9 | 3.2 | 2.7 | 181.5 |
| Organic | + | - | 3.3 | 2.8 | 2.4 | 137.5 |
| Organic | + | + | 5.5 | 3.8 | 3.5 | 157.0 |

Table 6. Soil microorganism population in the relation to different fertilizer system

The populations of soil microorganisms in the 6th radish culture were shown in Table 6. Manure input plots showed higher bacteria population than no manure input plots, but differences among fertilizer types was not clear. The populations of actinomyces and fungus were also not affected by treatment; however, the B/F value was high in the organic fertilizer plots manure input plots.

B/F value is one of the most important indicators for evaluating soil microorganism condition. Usually, high B/F values show healthier soil conditions than lower values (Kato et al, 1977). In this research, the population of soil microorganism was less than in farmer's fields, because this field had not been manured for many years. However, it was notable that organic fertilizer and manure input could improve soil microorganism conditions. These soil biological condition might enhance the radish growth.

### 3.2 Potato yield respond and related analysis in 1997

Potato growth and yields were shown in Table 7. At flowering season, above and underground biomass were higher in organic fertilizer plots than in chemical fertilizer plots irrespective of microbial materials application. While manure application was also effective in increasing above ground biomass yields in any fertilizer treatment, however, underground biomass decreased in manure plots. Potato yields were highest in organic fertilizer with manure application plots, followed by organic fertilizer with microbial materials and manure application plots.

Yano et al (1982) reported input of organic fertilizer improved potato yields significantly, in this research, manure input treatment showed high yields compared with no manure input plots. However, it is noteworthy that organic fertilizer also improved potato yields.

| Fertilizer | Microbial material | Manure | Flowering time (6 Jun) | | | | Yield |
|---|---|---|---|---|---|---|---|
| | | | Above ground biomass (g/plant) | Potato biomass (g/plant) | T/R ratio | Dry weight (g/plant) | (g/plant) |
| Chemical | - | - | 293.8 | 319.7 | 0.91 | 30.8 | 558 |
| Chemical | - | + | 451.2 | 304.1 | 1.83 | 41.7 | 770 |
| Organic | - | - | 353.5 | 482.3 | 0.74 | 38.5 | 822 |
| Organic | - | + | 458.4 | 282.5 | 1.71 | 39.8 | 945 |
| Organic | + | - | 518.3 | 568.4 | 0.91 | 54.1 | 706 |
| Organic | + | + | 848.3 | 583.2 | 1.46 | 76.9 | 869 |
| Significant | | | | | | | |
| Fertilizer type | | | * | * | N.S. | * | * |
| Manure | | | N.S. | N.S. | * | N.S. | * |

* and NS indicated significance at 5% and not significant.

Table 7. Potato growth and yields in the relation to different fertilizer system

| Fertilizer | Microbial material | Manure | Soil Nitarate (mg/100g dry soil) | | |
|---|---|---|---|---|---|
| | | | 30 Apr. | 23 May | 24 Jun |
| Chemical | - | - | 48.0 | 15.6 | 3.3 |
| Chemical | - | + | 81.0 | 21.3 | 3.6 |
| Organic | - | - | 12.3 | 12.3 | 2.7 |
| Organic | - | + | 51.0 | 14.7 | 3.0 |
| Organic | + | - | 60.0 | 11.4 | 3.9 |
| Organic | + | + | 135.0 | 10.8 | 4.2 |

Table 8. Nitrate concentration in the relation to different fertilizer system

The changing of soil nitrate concentration in potato culture in 1997 was shown in Table 8. On 30 April, organic fertilizer with microbial material plots showed the highest nitrate concentration, followed by chemical fertilizer plots, while organic fertilizer without microbial materials plots were lowest. Manure plots showed higher nitrate concentrations than no manure input plots. On 23 May, it was highest in chemical fertilizer plots, followed by organic fertilizer without microbial materials, while organic fertilizer with microbial materials plots were the lowest. However, these differences disappeared on 24 June. Microbial materials application enhanced nitrate release from organic fertilizer at the early stage of growth.

### 3.3 Economic productivity

The economic productivities between chemical and organic fertilizers for radish and potato culture from 1994 to 1997 were shown in Table 9. For no manure input plots, the cost of summer radish culture increased by 15 % for organic fertilizer, and 29% for organic fertilizer plus microorganism material compared with chemical fertilizer plots, so the cost per food weight also increased by 5% and 16%, respectively. The cost of autumn radish culture also increased 12% and 24%, respectively, however, so the cost per food weight decreased 4.3% for organic fertilizer plots, in spite of that increased 9% for the organic fertilizer plus microbial materials plot. In addition, the cost of potato culture increased 46% for organic fertilizer, and 74% for organic fertilizer plus microbial material compare with chemical fertilizer plots, however, so the cost per food weight decreased 5.4% for organic fertilizer plots , in spite of that increased 37.8% for organic fertilizer plus microbial materials plot.

For manure input plots, chemical fertilizer plots showed the lowest cost per food weight for summer radish, autumn radish and potato. For radish culture, manure input could not decrease the cost of food weight, but it did help to increase potato yields significantly.

These results suggested that the economic benefit of using organic fertilizer, microbial material, and manure differed depending on the crops. Organic fertilizer showed good performance for radish culture, while, chemical fertilizer plus manure input showed good performance for potato culture. In this research, the benefit of using microbial material was not clear, so, for the present results, microbial material may have some potential to improve crop yields, but their cost may hinder their use in crop production.

Organic fertilizer could be one of the most important materials to establish sustainable agriculture system. They can add organic matter in the soil and improve soil quality. In addition, to use organic fertilizer has more benefit for our society, because it can contribute to establish Recycle-Based Society. In this research, it was clear that organic fertilizer can improve radish and potato yields on a long term basis. However, another strategy to reduce the increasing cost to use organic fertilizer is needed.

### 3.4 Potato yield respond and related analysis in 2010

Soil fertility and crop quality could be improved by proper farming practices. Potatoes were planted in 2010, which including two different ridging managements. Soil properties and potato quality was observed in this study. The ridging heights were different because of different riding managements (Figure 1), which influenced the soil properties and thus influenced potato yield to some extent.

| Fertilizer | Microbial material | Manure | Summer radish | | | Autumn radish | | |
|---|---|---|---|---|---|---|---|---|
| | | | Yields | Cost | Cost | Yields | Cost | Cost |
| | | | kg ha-1 | yen ha-1 | yen 100kgFW | kg ha-1 | yen ha-1 | yen 100kgFW |
| Chemical | - | - | 526,700 | 1,640,000 | 3,110 | 546,600 | 1,940,000 | 3,553 |
| Chemical | - | + | 421,000 | 1,670,000 | 3,961 | 518,700 | 1,970,000 | 3,801 |
| Organic | - | - | 574,100 | 1,890,000 | 3,284 | 644,600 | 2,190,000 | 3,397 |
| Organic | - | + | 583,200 | 1,920,000 | 3,284 | 602,000 | 2,220,000 | 3,686 |
| Organic | + | - | 582,000 | 2,110,000 | 3,618 | 621,300 | 2,410,000 | 3,878 |
| Organic | + | + | 630,500 | 2,140,000 | 3,386 | 548,500 | 2,440,000 | 4,447 |

| Fertilizer | Microbial material | Manure | Potato | | |
|---|---|---|---|---|---|
| | | | Yields | Cost | Cost |
| | | | kg ha-1 | yen ha-1 | yen 100kgFW |
| Chemical | - | - | 306,900 | 630,000 | 2,049 |
| Chemical | - | + | 423,500 | 660,000 | 1,555 |
| Organic | - | - | 452,100 | 880,000 | 1,939 |
| Organic | - | + | 519,800 | 910,000 | 1,748 |
| Organic | + | - | 388,300 | 1,100,000 | 2,824 |
| Organic | + | + | 478,000 | 1,130,000 | 2,356 |

1) Seed, chemicals materials without fertilizer, power sources, machines, and labor costs were quated by vegetable production cost survey (MAFF, 1997).
2) The yields of summer and autumn radish were used in 6th and 5th culture. 2) The yields of summer and autumn radish were used in 6th and 5th culture.

Table 9. Economically productively in relation to different fertilizer system

Soil temperature was recorded every 30 minutes in the whole period of potato production in 2010. The results showed that soil average temperatures of Ridging 3 were 19.5°C and 19°C at 3cm and 10cm depth, and Ridging 2 were 18.8°C and 18.7°C, respectively (Figure 2). The highest soil temperature under Ridging 3 was 46.8°C at 3cm soil depth, and the highest soil temperature under Ridging 2 was 36.4°C at 3cm soil depth. The lowest temperature was 2.2°C under Ridging 3 and 2.0 °C under Ridging 2 at 3cm soil depth, respectively. The change of temperature of Ridging 3 treatment was more remarkable than Ridging 2 treatment.

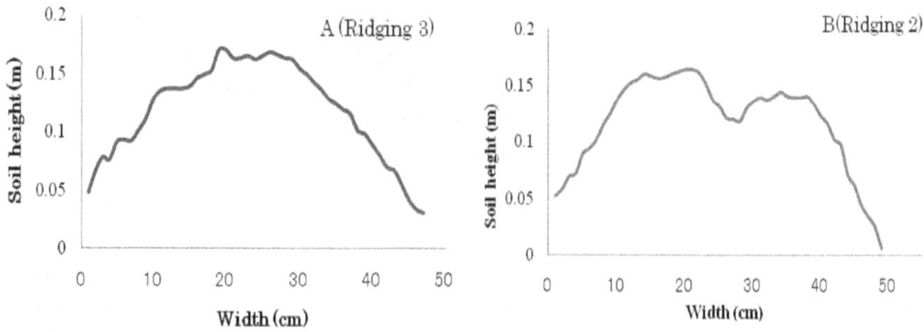


Fig. 1. The shapes of riding managements in potato production (2010). The ridging shapes of the two managements were measure on June 18, 2010.

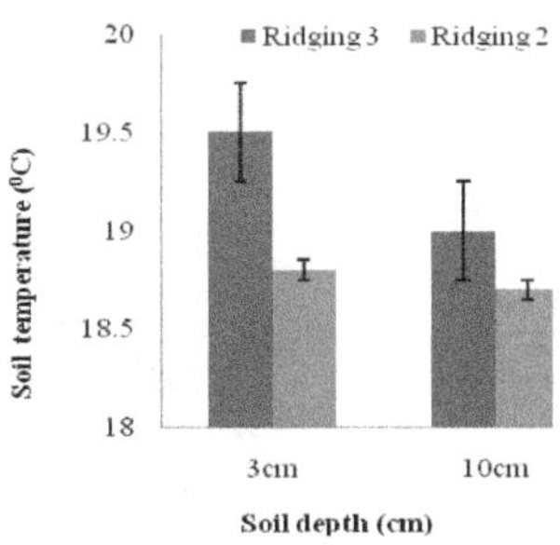


Fig. 2. Soil average temperatures in potato production (2010)

The effect of different ridging managements on soil PH and soil EC was insignificant, but the effect of dates on soil PH and soil EC was significant (Table 10). Though different ridging managements did not have any significant effect on soil PH and soil EC, but most soil PH and soil EC observed from Ridging 3 were higher than Ridging 2 in potato production. The effect of different ridging managements on soil nutrients concentrations was insignificant but the dates had remarkably significant effect on soil nutrients concentrations. Figure 3 show the soil nutrients concentrations in the potato production.

Table 11 and Table 12 showed the potato yields and characters in 2010. In flowering season, results showed that the effect of different ridging managements on potato root weights was significant, Ridging 3 promoted significantly the root growth than Ridging 2. Though different ridging managements did not have any significant effect on aboveground biomass, Ridging 3 produced the higher aboveground biomass than Ridging 2. The effect of different ridging managements on crop heights was insignificant and crop heights under the two managements were similar. The effect of different ridging managements on potato yield and number per plant in flowering season was insignificant and Ridging 3 produced a lower yield and number per plant than Ridging 2.

| Factor | PH | EC | Ca | K | Mg | $NH_4^+$-N | $NO_3^-$-N | P |
|---|---|---|---|---|---|---|---|---|
| Treatment | NS | NS | NS | NS | NS | NS | NS | NS |
| Date | *** | *** | *** | * | ** | *** | * | * |
| Treatment × Date | NS | NS | NS | NS | NS | NS | NS | NS |
| *, **, *** meant 5%, 1%, 0.1% significant each symbol. And NS meant not significant. | | | | | | | | |

Table 10. ANOVA of soil properties in potato production (2010)

At harvest stage, results showed that the effect of different ridging managements on yields and numbers of potato and marketable potato and potato per plant was insignificant. The yields obtained from Ridging 3 were lower than Ridging 2 for potato and marketable potato, however, the numbers obtained from Ridging 3 were higher than Ridging 2 for potato and marketable potato. The effect of different ridging managements on yields and numbers of potato and marketable potato was reversing. Ridging 3 produced a little bit lower potato yield per plant than Ridging 2 at harvest stage that was same as the results observed in flowering season, but the potato number per plant obtained from Ridging 3 was higher than Ridging 2 at harvest stage.

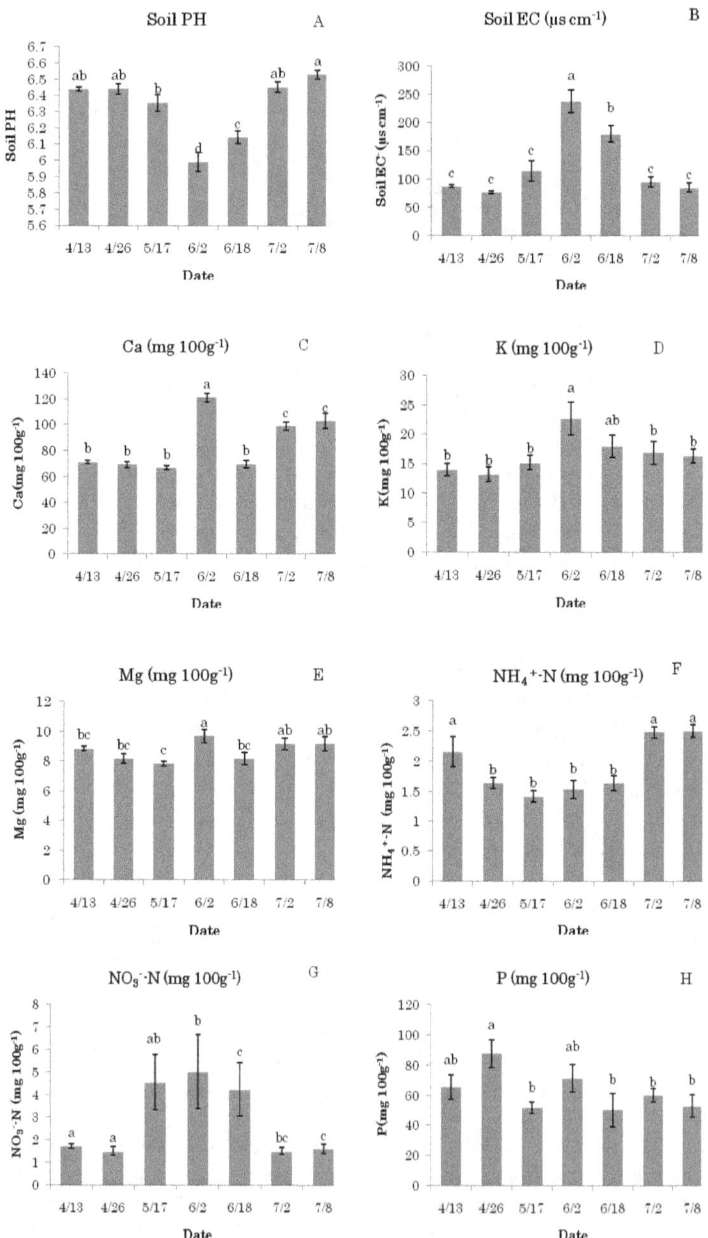


Fig. 3. Soil nutrients concentrations in potato production (2010). The effect of dates on soil PH, soil EC, soil Calcium, Potassium, Magnesium, $NH_4{}^+$-N, $NO_3{}^-$-N, and Phosphorus concentrations were significant. The alphabets indicated the difference between different dates (P≦0.05%).

Ridging 3 had the higher root weight of potato, but produced lower yield of potato, which may be explained that the soil temperatures were different under different ridging managements. The proper temperature of potato tuber growth is 20℃, and tuber growth stops if the temperature is over 30℃. Ridging 3 had higher soil temperatures, which promote the aboveground stem and leaf growth by photosynthesis, however, leading to less photosynthesis for tuber growth, so Ridging 3 had higher aboveground biomass than Riding 2 but produced lower yield of potato than Ridging 2. The number of potato increased and the weight of potato decreased if the soil temperatures were higher, so Ridging 3 produced higher potato number and lower average weight of potato than Riding 2.

Plate 1. Photos of potato production field in Ushiku City, Ibaraki Prefecture,Japan. In the flowering season, potato roots and shoots was collected (Photo A). In harvest stage, continuous ten plants was harvested mannually for potato yield and quality analysis (Photo B).

In this study, potato grade based on starch values was served as the evaluation factor. The starch values of high grade potatoes focused on 12%-19% (Yoshida, 1988). Average starch values and the percentages of high grade potatoes under two managements were shown in Table 13. Results showed that the effect of different ridging managements on potato starch values was significant. Ridging 2 produced higher average starch value, Ridging 3 produced more high grade potatoes. 87.7% potatoes obtained from Ridging 3 were high grade potatoes; however, 84.5% potatoes obtained from Ridging 2 were high grade potatoes. That suggested that Ridging 3 had positive effect on table potato production industry. There were 12.6% potatoes produced from Ridging 3 that starch values were less than 12%, however, there were 14.4% potatoes produced from Ridging 2 that starch values were less than 12%. 1.2% potatoes produced from Ridging 2 that starch values were over 19% and no potatoes produced from Ridging 3 that starch values were over 19% (Figure 4).

The weight of matured potato was served as the standard of evaluating appearance quality of potato in this research. Results showed that the effect of different ridging managements on potato weights was significant but on marketable potato weights was insignificant (Table 13). The average weight of potato obtained from Ridging 3 was lower than Ridging 2. The weight of marketable potato obtained from Ridging 3 was also lower than Ridging 2. Ridging 2 had positive effect on the weights of potatoes and marketable potatoes. 21.1% and 43.4% potatoes produced from Ridging 3 were S and M sizes, respectively, and 19.8% and 37.7 % potatoes produced from Ridging 2 were S and M sizes, respectively. Moreover, 24.0% and 2.3 % potatoes produced from Ridging 3 were L and 2L sizes， and 30.5% and 8.4% potatoes produced from Ridging 2 were L and 2L sizes (Figure 5).

| Treatment | Crop height (cm) | Aboveground biomass (g plant$^{-1}$) | Root weight (g plant$^{-1}$) | Potato yield (g plant$^{-1}$) | Potato number (plant$^{-1}$) |
|---|---|---|---|---|---|
| | Flowering season (June 2, 2010) | | | | |
| Ridging 3 | 45.6 | 24.3 | 13.8 | 112.8 | 4.7 |
| Ridging 2 | 46.7 | 18.7 | 5.0 | 118.5 | 5.3 |
| Significant | | | | | |
| Treatment | NS | NS | * | NS | NS |
| * meant 5% significant symbol. And NS meant not significant. | | | | | |

Table 11. Potato growth and yields in flowering season

| Treatment | Yield (g plant$^{-1}$) | Number (plant$^{-1}$) | Yield (t ha$^{-1}$) | Number (000 ha$^{-1}$) | Marketable Yield (t ha$^{-1}$) | Marketable Number (000 ha$^{-1}$) |
|---|---|---|---|---|---|---|
| | Harvest stage (July 8, 2010) | | | | | |
| Ridging 3 | 590.7 | 5.8 | 31.4 | 308.7 | 31.0 | 289.4 |
| Ridging 2 | 635.5 | 5.6 | 33.5 | 293.4 | 32.9 | 284.5 |
| | | | | Significant | | |
| Treatment | NS | NS | NS | NS | NS | NS |
| NS meant not significant. | | | | | | |

Table 12. Potato growth and yields at harvest time

| Treatment | Average weight of marketable potato (g) | Average weight of potato (g) | Average Starch value of potato (%) | High grade potatoes (%) | Marketable and high grade potatoes (%) |
|---|---|---|---|---|---|
| Ridging 3 | 107.7 | 101.3 | 13.3 | 87.7 | 84.6 |
| Ridging 2 | 115.1 | 114.2 | 13.8 | 84.5 | 82.6 |
| | | | Significant | | |
| Treatment | NS | * | * | NS | NS |
| * meant 5% significant symbol. And NS meant not significant. | | | | | |

Table 13. Potato quality in 2010

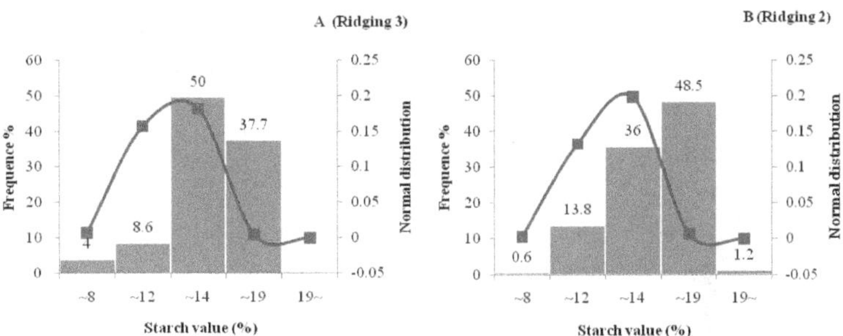


Fig. 4. Histogram and normal distribution of starch values of potatoes (2010). Most starch values obtained from Ridging 3 focused on 13%~14% (A), most starch values obtained from Ridging 2 focused on 15%~19% (B).

The starch values of high grade potatoes were in the range of 12%~19%, and the weights of potatoes were over 30g recording as marketable potatoes. The marketable and high grade potatoes were the potatoes that starch values and weights met above two conditions. The percentage of marketable and high grade potatoes produced from Ridging 3 was 84.6% and that produced from Ridging 2 was 82.6% (Table 13 ). Ridging 3 produced more marketable and high grade potatoes than Ridging 2. The relationships between starch values and weights of marketable and high grade potatoes were shown in Figure 6.

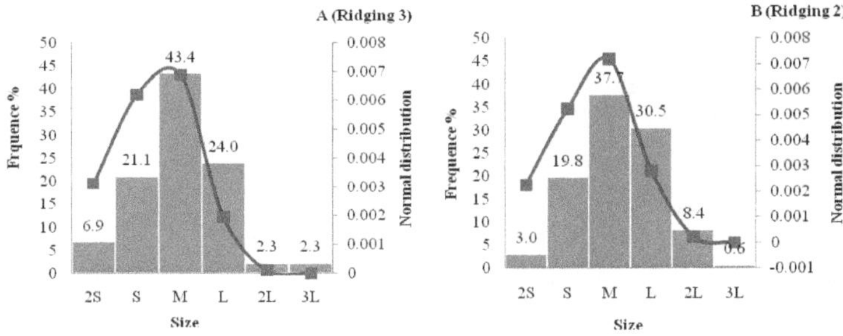


Fig. 5. Histogram and normal distribution of weights of potatoes (2010). Ridging 3 had positive effect on S and M sizes potatoes (A), and Ridging 2 had positive effect on L and 2L sizes potatoes (B).

These results suggested that the 3 time ridging promote the appropriate shape of the ridge that enhances the soil temperature in the ridge. These effects also enhance the roots distribution in the ridge, and improve the ratio of high grade potato compare with 2 time ridging. These differences may be influenced by differences decompose speed of organic fertilizer *Bochashi* between 2 and 3 time ridging treatments.

### 3.5 Conclusion

Chemical fertilizers consumption in agriculture sector reached 5 times level of 1975 in 1990 and increased slightly afterwards in Japan. In Asian countries, since Jump in crude oil prices in 21st century, as increase chemical fertilizer price due to increasing oil prices, most farmer

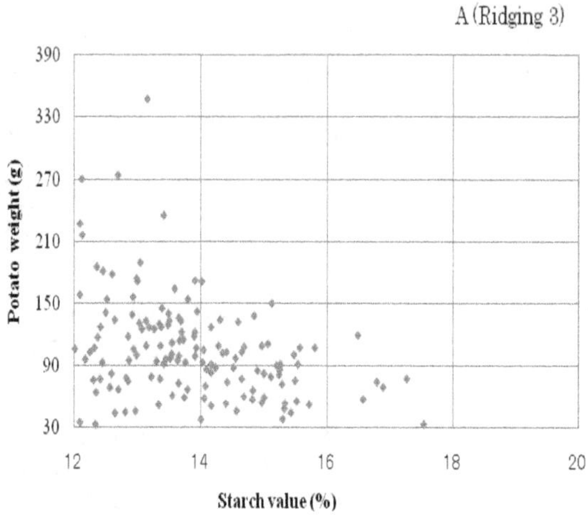


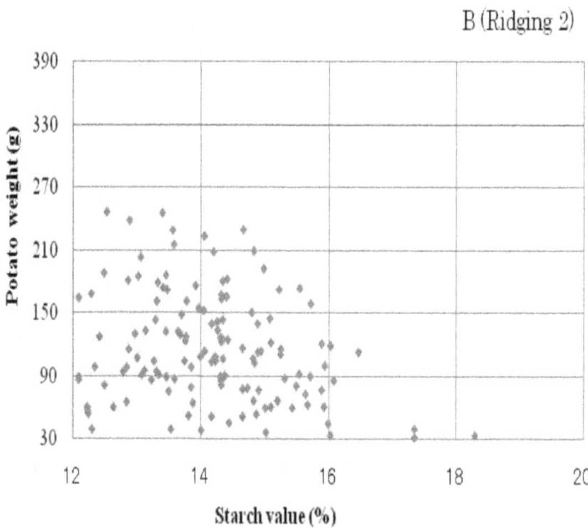


Fig. 6. The relationships between starch values and weights of marketable and high grade potatoes (2010).

favorable to use organic fertilizer. *Bocashi* is the traditional way to make compost using agricultural sub products and waste in Japan, and this technique is widely spared to the other Asian countries. In this chapter, we discussed the benefits of organic fertilizer for radish and potato production.

There were no significant difference between chemical and organic systems for 1 or 2 years, however, it 1was clear that the radish growth in organic fertilizer plots was higher than in chemical fertilizer plots 2 or 3 years later. Potato yields were highest in organic fertilizer with manure application plots, followed by organic fertilizer with bacteria materials and manure application plots. The lowest value was in unmanured chemical fertilizer plots. Manured plots showed higher bacteria populations than unmanured plots, however, difference among fertilizer types were not clear. The populations of actinomyces and fungus were also not affected treatment, however, the B/F value was high in the organic fertilizer plots and manured plots. While organic fertilizer showed good performance for radish culture, chemical fertilizer plus manured showed good performance for potato culture. The benefit of using microbial material was not clear, so, for the present results, microbial material may have some potential to improve crop yields, but their cost may hinder their use in crop production.

The use of *bocashi* organic fertilizer as alternative soil fertility amendments in nutrients effective to eliminate the chemical fertilizer, enhance the sub materials from rice production.

We also discussed the appropriate farming practice based organic fertilizer application management for potato production in farmer's field. The results suggested that 3 time ridging promote the appropriate shape of the ridge that enhances the soil temperature in the ridge. These effects also enhance the roots distribution in the ridge, and improve the ratio of high grade potato compare with 2 time ridging. These differences may be influenced by differences decompose speed of organic fertilizer *Bocashi* between 2 and 3 time ridging treatments. These results would contribute the development for conservation farming system for eliminating the use of chemical fertilizer.

## 4. Acknowledgment

We appreciate Iseki Kanto Co. Ltd. for their kind assistance to supply the experimental materials. We appreciate Mr. Motomu Takamatsu and Mr. Shingo Abe for their technical assistance to provide the field experiment. This work was supported by Grant-in-Aid for Scientific Research (C) (22580286).

## 5. References

Brian Baker, Sean L. Swezey ,David Granatstein, Steve Guldan, David Chaney (2005). Organic Farming Compliance Handbook: A Resource Guide for Western Region Agricultural Professionals, Western Region USDA SARE program. P 1-3.

Carter, M. R., Holmstrom, D., Sanderson, J. B., Ivany, J. and DeHaan, R. (2005) Comparison of conservation with conventional tillage for potato production in Atlantic Canada: crop productivity, soil physical properties and weed control. Can. J.Soil Sci. 85: 453–461.

Jongtae Lee (2010). Effect of application methods of organic fertilizer on growth, soil chemical properties and microbial densities in organic bulb onion production. Scientia Horticulturae, 124, 299–305.

Matsumoto, M (1983) roots growth, Nogyogizyutu taikei, Vegetable 9, radish, carrot and turnip,23-40, Noubunkyo, Tokyo.

Noguchi, K., (1992a) Organic fertilizer and soil micro livings 1, Nogyo oyobi Engei. 67:673-677.

Noguchi, K., (1992b) Organic fertilizer and soil micro livings 3, Nogyo oyobi Engei. 67:883-887.

Kato, K., E. Kokawa, S. Tsuru, and T. Suzuki (1977) Bacteria-Fungi ratio in several soils, Journal of the science of soil and manure, Japan, 48(9,10):437-438.

Yoshida(1988). Encyclopedia of Potato. Rural Culture Association Japan. P 77.

# Part 2

# Understanding and Making Decisions

5

# Understanding Sugarcane Yield Gap and Bettering Crop Management Through Crop Production Efficiency

Fábio Ricardo Marin

*Scientific Researcher, Embrapa Agriculture Informatics, Campinas, SP Brazil*

## 1. Introduction

The comparison among farming systems and regions would improve the understanding of how and what driving factors explains the crop yield variability over time and space. Very often, however, farm managers and policy makers fall in difficult to establish reliable indexes to compare farming systems plots and regions. Having a quantitate index, we could derive relationships regarding climate, soil and socioeconomic, as well as to determine which factors contribute or hinder the development in a given region and time.

Monteith (1977) suggested agroecosystems as machines that utilize solar energy to maintain composition and organization. From a thermodynamic standpoint, the efficiency of any process can be expressed as the ratio of energy output to energy input. Since the 1970s, this concept has been applied to analyze the energy flow in agroecosystems, as well as to analyze the relation between biomass chemical energy and incident solar radiation.

We could apply this approach to understand the regional agricultural development and crop yield gap, once it could elucidate biophysical factors, such as the pedoclimatic conditions, affecting crop yields at a local scale. However, for a broader evaluation, one should also include structural components, corresponding to the agricultural systems and management practices adopted; institutional effects, involving governmental actions affecting price, credit, commercialization, and incentives; and research and development, related to innovations to increase yield and solve problems that restrict agricultural-related activities (Carvalho, 2009).

Also, to make this approach useful in an operational way, one could assume crop efficiency such as a quantitative indicator, helping to compare and evaluate in time and space, the farming development level. The efficiency of crop production can be assumed as the ratio between observed and attainable crop yield (Marin et al., 2008).

In order to evaluate the effectiveness of this tool, the concept of crop efficiency was applied to study the sugarcane performance in the State of São Paulo, Brazil, the main region of this crop production, representing approximately 60% of the total Country's sugarcane production (IBGE, 2002).

## 2. Methods and input data

The weather data had been supplied by the Brazilian Agrometeorological Monitoring System (EMBRAPA INFORMÁTICA AGROPECUÁRIA, 2002), comprising the period between 1990 and 2006. The weather data was organized in a 10 day time step. Daily solar radiation values were simulated using the Bristow and Campbell (1984) method previously calibrated using A=0.7812, B=0.00515, and C=2.2 as model parameters.

An empirical model derived from Doorembos & Kassan (1979) was used to assess the potential (PY) (Equation 1) and attainable water limited yield (WLY) as proposed by Jensen (1968) (Equation 2).

$$PY = -6.2501 + 0.2187\ S + 0.3304\ T \quad (\text{t ha}^{-1}\ 10\ \text{day}^{-1}) \tag{1}$$

where T is the mean air temperature (°C) for 10 days and S is the incident solar radiation (MJ $m^{-2}\ d^{-1}$).

$$\frac{WLY}{PY} = \left(\frac{ETa_1}{ETm_1}\right)^{\lambda_1}\left(\frac{ETa_2}{ETm_2}\right)^{\lambda_2}\left(\frac{ETa_3}{ETm_3}\right)^{\lambda_3} \tag{2}$$

where $\lambda_1$=0.43, $\lambda_2$=0.39, and $\lambda_3$=0.07 are water deficit sensibility factors for each of the three crop phases, such as: 1) initial, from planting to 30 days after (DAP), 2) crop development, up to 330 DAP, and 3) late, up to 365 DAP.

The actual crop evapotranspiration (ETa) was computed for a 10 day time step using a simple crop water-balance simulation (Thornthwaite & Mather, 1955). The Kc coefficients and development stages used were described by Doorembos & Kassan (1979) (Table 1) and available soil water was chosen according to Smith et al. (2005). Reference evapotranspiration (ETo) was estimated following Camargo et al. (1999), which was modified from Thornthwaite (1948) to match with Penman-Monteith method (Allen et al., 1998) using just air temperatures as input weather data.

Crop coefficients were obtained in Doorembos & Kassan (1979) by assuming a 12 months growing cycle, using the adjustments provided by Barbieri (1993). The simulations were done for three growing seasons (May to April, July to June, and October to September) representing the typical ratoon crop in early , middle and late growing seasons. The results from each year were averaged, and the average was used as a reference yield to efficiency calculation.

Actual sugarcane yield values (AY) for each county of the São Paulo State during the growing seasons of 1990-1991 and 2005-2006 were obtained from the Brazilian Institute of Geography and Statistics (IBGE) (www.sidra.ibge.br). Both AY and WLY dataset were spatially organized and their maps were generated by ordinary kriging interpolation tool in ArcGIS 9.3 (ESRI, Redlands, CA), using a 900 m spatial resolution grid.

The soil fertility was taken into account in the empirical model through a soil correction factor (SCF) varying from 0,74 to 1 (Table 1) based on Prado (2005), who classified the soils State of Sao Paulo considering their suitability for sugarcane production. In that classification, Prado (2005) states four yield ranges for sugarcane. The values presented in Table 1 mean the normalized values of those yield ranges. To apply this concept for the State of Sao Paulo, the soil map of the State was re-classified using the criteria suggested by Prado

(2005). Using the raster calculator tool available in ArcGIS 9.3 (ESRI, Redlands, CA), the SCF maps were multiplied by WLY maps to produce a map of soil and water limited yield (SWLY) for every growing season.

| Aptitude | soil correction factor |
|---|---|
| Good | 1.00 |
| Regular | 0.94 |
| Restrict | 0.84 |
| Inadequate | 0.74 |

Table 1. Soil correction factor (SCF) for sugarcane in the State of São Paulo.

In order to obtain the sugarcane efficiency maps for the State of Sao Paulo, those AY maps were divided by AY maps using the raster calculator tool in ArcGIS 9.3 (ESRI, Redlands, CA). This procedure had been repeated for every season, resulting in a set of 16 efficiency maps.

To quantify the soil and sugarcane production efficiency (SPE) relationship, soil aptitude classes were converted into a numerical rank from 1 to 4 and the Spearman Rank Correlation (SRC) coefficient (Snedecor & Cochran, 1982) was applied. To correlate efficiency with the others variables - air temperatures, rainfall, water deficit and solar radiation- the Pearson method (PC) was used (SNEDECOR; COCHRAN, 1982). Socio-economic (SE) influences on SPE, as well as the influence of crop management (varieties, diseases, pests etc.) was assumed to be the complimentary value to the sum of correlation indexes regarding soil and climate variables (Equation 3).

$$SE = 1 - SRC - PC \tag{3}$$

## 3. Sugarcane crop efficiency in the state of Sao Paulo

Sugarcane is one of the world's major food-producing C4 crops, providing about 75% of world sugar harvested for human consumption (Souza et al., 2008) and one of the most important crops for the Brazilian economy. More recently, sugarcane has also become recognized as one of the central plant species for energy production as liquid fuel and electricity (Goldenberg, 2007). Biofuels are, at present, the fourth source of primary energy after oil, coal and gas. Brazil is the world's largest exporter of ethanol and the world's second largest producer, the US being the largest. In 2006 Brazil alone produced 16.3 billion liters, 33.3% of the world's total ethanol production and 42% of the world's ethanol used as fuel, and from then on ethanol production increased from year to year. Particularly in the US, Brazil, the EU and some Asian countries, government-led incentive programs focus on renewable sources of energy. The main driver behind these recent efforts to increase the volume of biofuels in the energy mix are concerns over climate change and greenhouse gas (GHG) emissions (primarily $CO_2$), and widely fluctuating oil prices with the desire to diversify and stabilize energy supplies. In addition to the commercial uses for sugar, ethanol and electricity in mills, the crop is widely used by small farmers around the country as feedstock for animals or as raw material for homemade rum and brown sugar.

The overall SPE average for the State of Sao Paulo was 48%, increasing from 0.42 to 0.57 throughout the analyzed period. Between 1990/1991 and 1995/1996, the SPE oscillated around 0.45 as a result of the tough macroeconomic conjuncture experienced by Brazil at that period, as well as due the unfavorable conditions for sugar and ethanol commercialization (GOLDEMBERG et al., 2007). However, an expressive yield increase has occurred in the last 6 years of the time series (Figure 2), as a result of the increased ethanol consumption in Brazil. This, in turn, was a consequence of better gasoline-ethanol price ratio since the beginning of the 2000s, and the availability of bi-fuel vehicles in Brazil after 2002 (Macedo, 2007).

Along the analyzed period, the average yield of the State of Sao Paulo increased 12 t $ha^{-1}$ (Figure 2). Based on this, we derived that for each SPE percentage point increased there was a rise of 0.8 t $ha^{-1}$. Extrapolating it for the current sugarcane growing area in the State of Sao Paulo, it would represent an increase of 2 million tons of cane per each percentage point. This number takes especial importance when discussing the expansion of Brazilian sugarcane growing area (Manzatto et al., 2009), meaning that by driving new investments to zones with higher SPE, less land would be needed to supply the Brazilian and international sugar and ethanol demands.

The SPE maps showed northern and central region as the areas where SPE had the higher increase rates as a consequence of the new mills installed in those regions during this decade (Figure 3). High SPE areas (>80%) showed the higher expansion along the time (Figure 3d and 3e), while low SPE (<20%) were reduced in about 30% (Table 3, Figure 3b e 3c).

Areas with SPE higher than 80% expanded from 17610 $km^2$ to 68754 $km^2$ (Table 3), denoting the intensification of land use in the State of Sao Paulo and new production pattern in sugarcane fields. In the traditional areas growing sugarcane, where SPE is normally higher, this process may be a consequence of the use of better crop management mainly through varieties, fertilizers, and harvest management (Figure 3k, 3o and 3p).

In the newer areas, where SPE is lower, the SPE increase seems to be a consequence of the replacement of non-commercial sugarcane areas, used for animal feeding and home uses, by the commercial ones (sugar mills oriented), as sugar mills had expanded to those regions and had incorporated an important land amount to the sugarcane production system. This occurred mostly after 2002 and the SPE increasing trend seems to be a consequence of the investments applied to get suitable lands for sugarcane production.

In order to identify the relative importance of SPE drivers, we found climate as responding for 43% of spatial variability of SPE, while soil responded for 15% (varying from 10% to 18%) of the SPE variability, as an overall average across spatial and time scales. Therefore, the soil plus climate related factors responded for 58% of total SPE variability (Table 4), from which we derived that biotic, management and socio-economic factors together explained up to 42% of SPE variability.

Breaking the climate determination coefficient up into its components, we found solar radiation as the most important factor, followed by water deficit, maximum temperature, rainfall and minimum temperature (Table 5). Solar radiation as the higher determination coefficient variable may be due to the fact of most of the sugarcane growing areas have occupied some of the best agricultural areas of the State of Sao Paulo, where yield limiting factors have less influence. Thus, the crop was able to respond to a potential yield related variable, such as solar radiation (Bowen & Baethgen, 2002). In spite the inclusion of new areas at the west of the State of Sao Paulo, this has occurred just in the last few years, minimizing its impact into the analysis.

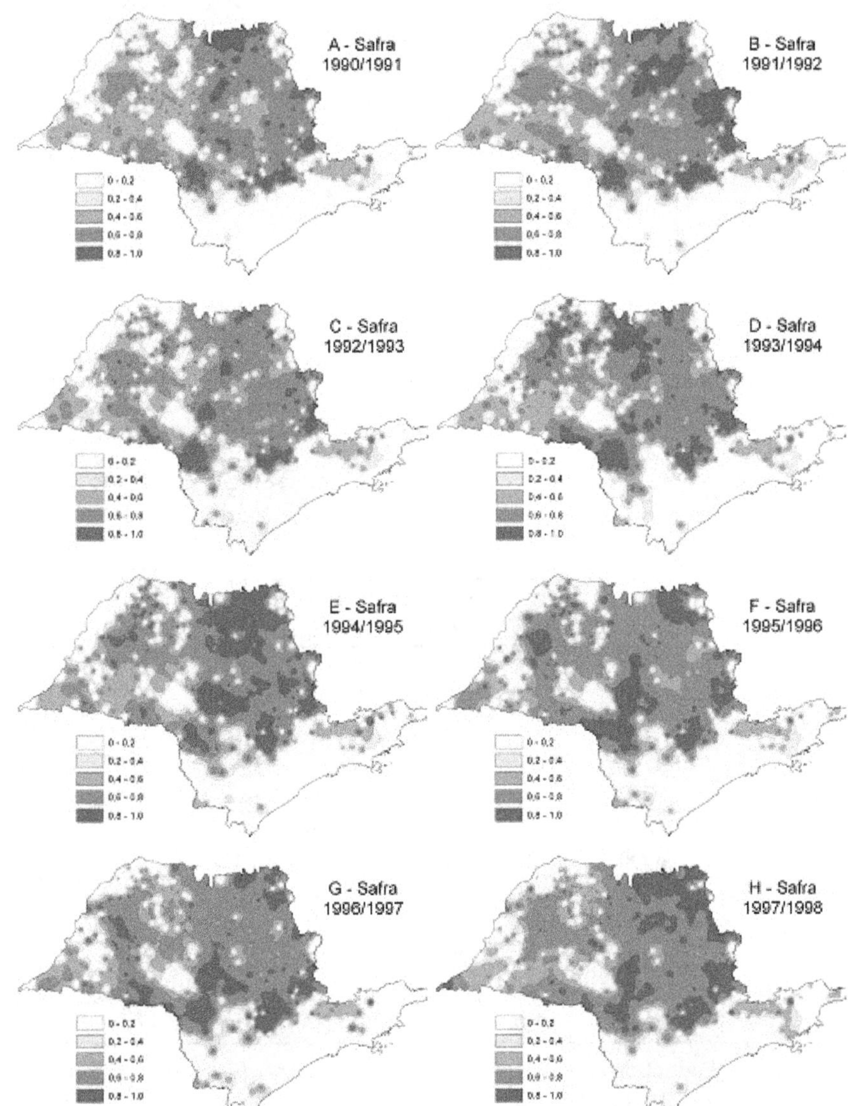


Fig. 1. Sugarcane production efficiency in the State of Sao Paulo from season 1990/1991 to 1997/1998.

Water deficit explained 12% of SPE variability, once rainfall amount and distribution seems to be enough to assure certain levels of sugarcane yield even in the worst years along the time series herein analyzed. Even in the western of Sao Paulo, where water deficit usually gets higher than other regions, the sugarcane yield still variation within a high yield range. However, we may infer that the same analysis including higher water deficit locations would certainly result in a higher $R^2$ for water deficit.

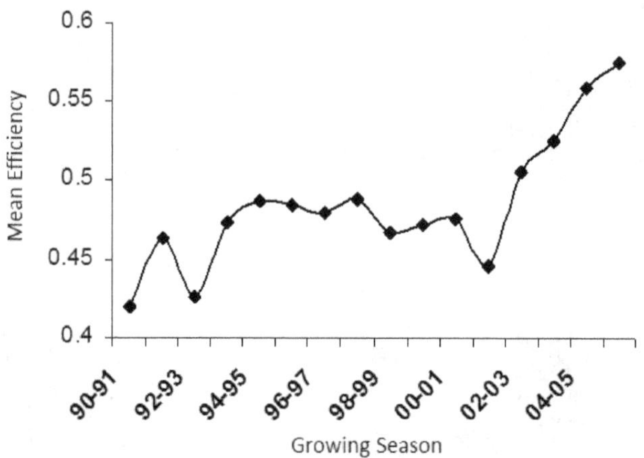


Fig. 2. Time variation of sugarcane production efficiency in the State of Sao Paulo between seasons 1990/1991 and 2005/2006.

| **Efficiency Class** | **Growing Season** | | | | | | | |
|---|---|---|---|---|---|---|---|---|
| | **90/91** | **91/92** | **92/93** | **93/94** | **94/95** | **95/96** | **96/97** | **97/98** |
| **0 - 0.2** | 74369 | 68163 | 75092 | 67202 | 65013 | 65226 | 67059 | 66000 |
| **0.2 - 0.4** | 31141 | 28129 | 32060 | 31230 | 29030 | 29086 | 26558 | 24244 |
| **0.4 - 0.6** | 60006 | 50706 | 48556 | 44223 | 42757 | 35303 | 37199 | 38015 |
| **0.6 - 0.8** | 65083 | 72108 | 76701 | 76103 | 67343 | 85220 | 86557 | 81450 |
| **0.8 - 1.0** | 17611 | 29103 | 15801 | 29451 | 44066 | 33374 | 30836 | 38501 |
| **Efficiency Class** | **Growing Season** | | | | | | | |
| | **98/99** | **99/00** | **00/01** | **01/02** | **02/03** | **03/04** | **04/05** | **05/06** |
| **0 - 0.2** | 67847 | 69055 | 76600 | 77018 | 66247 | 64055 | 60851 | 57088 |
| **0.2 - 0.4** | 23441 | 24970 | 25011 | 28593 | 23059 | 23037 | 19050 | 18367 |
| **0.4 - 0.6** | 47039 | 36473 | 31311 | 36012 | 39253 | 33010 | 23628 | 23083 |
| **0.6 - 0.8** | 90020 | 87013 | 72275 | 86085 | 80225 | 77207 | 79900 | 80916 |
| **0.8 - 1.0** | 19862 | 30699 | 43012 | 20501 | 39425 | 50900 | 64780 | 68754 |

Table 3. Sugarcane production efficiency area distribution classes ($km^2$) in the State of Sao Paulo from season 1990/1991 to 2005/2006.

| Driver factor | Determination Coefficient |
|---|---|
| Climate | 0.43 |
| Soil | 0.15 |
| Total | 0.58 |

Table 4. Average determination coefficients between climate and soil variables with sugarcane production efficiency in the State of Sao Paulo from season 1990/1991 to 2006/2007.

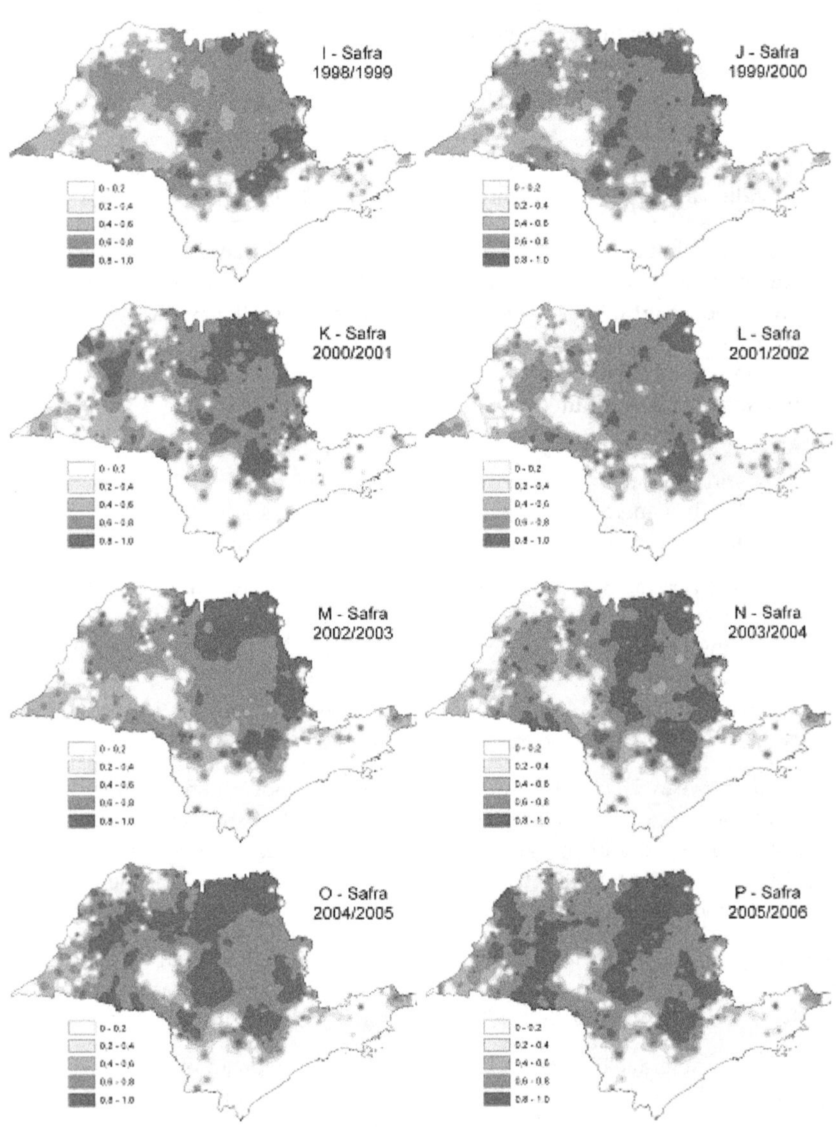


Fig. 3. Sugarcane production efficiency in the State of Sao Paulo from season 1998/1999 to 2005/2006 (continuation).

| Variable | $R^2$ |
|---|---|
| Solar radiation | 0.16 |
| Water deficit | 0.12 |
| Maximum temperature | 0.08 |
| Rainfall | 0.06 |
| Minimum temperature | 0.01 |
| Total | 0.43 |

Table 5. Average Pearson coefficient ($R^2$) from season 1990/1991 to 2005/2006, for solar radiation, rainfall, water deficit, maximum and minimum temperature.

The aggregation of climatic data into 10 day time step should be also considered as it has reduced time variability associated climatic variables. Also, matters to remember that analysis were based on growing season time-step average, and this really eliminated almost temporal variability. Thus, additional to the reasons discussed for water deficits, the results obtained for rainfall and temperatures seem to be related with data aggregation, as most of the time variability has been diluted by averaging the values over time.

The remaining 42% explaining the non-abiotic SPE drivers may be time-related to public policies, prices, and costs, mainly. Management and genetic improvements are also included in the amount, but in general the signals due to such factors are better expressed using a constant increasing rates, rather than a variable cause affecting yields.

By comparing fertilizer consumed and the Spearman index we intended to explore the effect of soil management on SPE (Figure 4). For this evaluation we hypothesized that seasons under tough economic conditions for growers should show higher correlation between soil and SPE. In opposite, when economy had been favorable to sugarcane business, lesser correlation between soil and SPE would be expect, since the fertilizer application reduced the fertility deficiencies in poorer soils, masking soil spatial variability.

For the period between 2002/2003 and 2005/2006, both Spearman and consumption of fertilizers have increased, contradicting the hypothesis just postulated. It may be due the intensive expansion of sugarcane growing areas to the west of the State of Sao Paulo, occupying less fertile soils than the traditional areas and thus increasing the importance of soil to explain SPE variability.

Thus, assuming that the hypothesis addressed before as correct, we can expect the SPE-soil correlation to fall in the coming years, since new the soil fertility of those new areas would be gradually improved over time, as can be observed after 2004 (Figure 4).

Since 2004, the State average observed yield reached 50 t $ha^{-1}$ spread over a wider area of the State of Sao Paulo. At the same time, the average attainable yield was 93 t $ha^{-1}$ in such a way that SPE was 0.52 in 2003/2004, 0.56 in 2004/2005 and in 0.57 during 2005/2006 growing season. At the same time, sugar price rose from US$ 11.3/50kg to US$ 20/50kg in just 1 year, seems to be related to that strong increase in SPE. The sugar price-SPE correlation analysis resulted in $R^2$=0.53 showing a high influence of the commodity prices to explain the SPE. Interesting to remember sugar prices being self-correlated with climate variables in Brazil, as Brazil in the largest producer in the world, and that is why sugar price-SPE $R^2$ had a value higher than 0.42 as it should be expected.

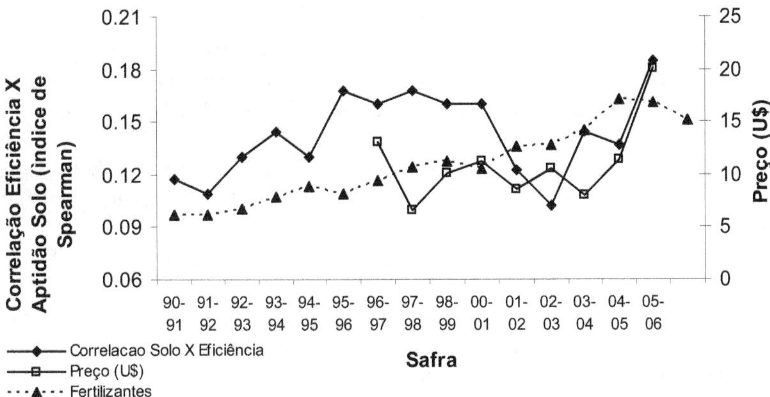


Fig. 4. Sugar prices (U$ per 50kg), commercialized amount of fertilizers in Brazilian central region ($10^9$ tons) after Ferreira & Gonçalves (2007) and average sugarcane production efficiency in the State of Sao Paulo.

## 4. Conclusion remarks

The sugarcane crop efficiency increased from 0.42 to 0.58 throughout the period from 1990 to 2006. The efficiency class above 80% showed the higher increase rates along that period. The crop yield gap has been reduced from 58% to 42%, possibly indicating the effect of the adoption of new technologies and the expansion of new mills in the west of the State of São Paulo.

The main abiotic variable explaining the sugarcane crop efficiency was the solar radiation (R2=0.16). All climate elements together explained nearly 43% of SEP variability. In average, 15% of SEP variability was explained to soil variability, with two different patterns: one from 1990 to 2001 and another from 2002 to 2006.

Adding climate and soil factors, we got biotic factors explaining 58 of SEP variability. It implies that 42% of SEP variability were explained by others factors, such as sugar prices.

## 5. Acknowledgements

This research was partially supported by Brazilian Council for Scientific and Technological Development (CNPq) through the projects 478744/2008-0 and 0303417/2009-9.

## 6. References

Allen, R. G, L. S Pereira, D. Raes, M. Smith. 1998. Crop evapotranspiration-Guidelines for computing crop water requirements-FAO Irrigation and drainage paper 56. FAO, Rome 300.

Barbieri, V. 1993. Condicionamento climático da produtividade potencial da cana-de-açúcar (*Saccharum spp*): um modelo matemático-fisiológico de estimativa. 1993. 142 p. Tese

(Doutorado em Fitotecnia) - Escola Superior de Agricultura "Luiz de Queiroz", Universidade de São Paulo, Piracicaba.

Camargo, A.P.; Marin, F.R.; Sentelhas, P.C.; Picini, A.G. 1999. Ajuste da equação de Thornthwaite para estimar a evapotranspiração potencial em climas áridos e superúmidos, com base na amplitude térmica. *Revista Brasileira de Agrometeorologia*, Santa Maria, v. 7, n. 2, p. 251-257.

Carvalho, L.G. 2009. Sugarcane crop efficiency in the State of São Paulo between growing seasons of 1990/ 1991 and 2005/2006. MSc dissertation. University of São Paulo. Brazil. 118p.

CONAB. Disponível em: http://www.conab.gov.br/conabweb/download/safra/cana.pdf. Acesso em: 29 maio 2008

Conceição, M.A.F.; Marin, F.R. 2007. Avaliação de modelos para a estimativa de valores diários de radiação solar global com base na temperatura do ar. *Revista Brasileira de Agrometeorologia*, Santa Maria, v. 15, n. 1, p.103-108.

Doorembos, J.; Kassan, A.H. 1994. Efeitos da água no rendimento das culturas. Roma: FAO, 212 p. (Estudos FAO: Irrigação e Drenagem, 33).

Embrapa Informática Agropecuária. *Agritempo*: sistema de monitoramento agrometeorológico. Campinas: Embrapa Informática Agropecuária; Cepagri/Unicamp, 2002. Disponível em: <http://www.agritempo.gov.br>. Acesso em: 05 jun. 2008.

Ferreira, R.; Gonçalves, J.S. 2007. Evolução e sazonalidade do consumo de fertilizantes no Brasil e nas unidades da federação no período 1987-2005. *Informações Econômicas*, São Paulo, v. 37, n. 11, p. 33-39.

Goldemberg, J. 2007.Ethanol for a sustainable energy future. *Science*, Washington, v. 315, p. 808-810.

Goldemberg, J.; Lucon, O. 2007. Energia e meio ambiente no Brasil. *Estudos Avançados*, São Paulo, v. 21, n. 59, p. 7-20.

Instituto Brasileiro De Geografia E Estatística. 2008. *Produção agrícola municipal*. 2002. Disponível: <http://www.ibge.gov.br/home/estatistica/economia/pam/2002/default.shtm>. Accessado em: 30 Jun.

Jensen, M. E. 1968. *Water consumptions by agricultural plant growth*. New York: Academic, v. 2.

Macedo, I.C. 2007.Situação atual e perspectivas do etanol. *Estudos Avançados*, São Paulo, v. 21, n. 59, p. 157-165.

Manzato, C. 2009. Zoneamento agroecológico da cana-de-açucar. Rio de Janeiro: Embrapa Solos.

Marin, F.R.; Lopes-Assad, M.L.; Assad, E.D. ; Vian, C.E.; Santos, M.C. 2008. Sugarcane crop efficiency in two growing seasons in São Paulo State, Brazil. *Pesquisa Agropecuária Brasileira*, Brasília, v. 43, p. 1449-1455,

Prado, H. 2005. Ambientes de produção da cana-de-açúcar na Região Centro-Sul do Brasil. *informações Agronômicas*, Cidade, n. 110, p. 6-9,

Richardson, C.W.; Wright, D.A. 2005. WGEN: a model for generating daily weather variables. Washington: USDA, *Agriculture Research Service*, 1984. 83 p. (ARS, 8).

Smith, D.M.; Inman-Bamber, N. G., Thorburn, P. J. Growth and function of the sugarcane root system. *Field Crop Research*, Amsterdam,

Snedecor, G.; Cochran, W.G. 1982. *Statistical methods*. 7th ed. Ames: Iowa State University Press, 507 p.

Thornthwaite, C.W.; Mather, J.R. 1955. *The water balance*. New Jersey: Drexel Institute of Technology, 104 p. (Publications in Climatology, n.8).

6

# A Conceptual Model to Design Prototypes of Crop Management: A Way to Improve Organic Winter Oilseed Rape Performance in Farmers' Fields

Muriel Valantin-Morison[1*] and Jean-Marc Meynard[2]
[1]*UMR 211 d'Agronomie INRA-AgroParisTech*
*Bât. EGER - BP01, F-78850 Thiverval-Grignon*
[2]*INRA Département SAD – Bât. EGER - BP01 F-78850 Thiverval-Grignon*
*France*

## 1. Introduction

Concerns about the adverse impacts of pesticides on the environment and their inevitable negative side-effects on non-target organisms have been growing since the 1960's. In western Europe, Winter Oilseed Rape (*Brassica napus* L. - WOSR) has to cope with numerous damaging insects (Alford et al., 2003 and Williams et al., 2010), and other diseases and weeds. In conventional farming, as a consequence, an important and increasing use of pesticides is observed in the major french production areas (Ecophyto R&D, 2009, Schott et al, 2010).For the same reasons, WOSR is not widely used in French organic farming as its yield is often low and unpredictable. Nevertheless, this "break" crop is of potential value in terms of market requirements, all the more that stakeholders need regular production to match with the organic oil demand from consumers. The agronomic benefit for the following wheat crop (Kirkegaard *et al.*, 1994), and the efficiency of winter oilseed rape as a nitrate trap crop (Vos and Vander Putten, 1997; Dejoux *et al.*,2003) are also well known.

In order to improve the performance of this crop in organic system, the crop management has to combine cultural, biological and mechanical way to reduce pests and diseases. Moreover, many scientists have been arguing, for more than two decades, that the reliance on chemicals could be considerably reduced by making better use of biotic interactions between pests, pathogens, weeds, crop and cultural practices (Chauvel et al., 2001, Meynard et al, 2003, Shennan et al., 2008, Aubertot et al., 2006, El-Khoury and Makkouk , 2010, Lucas , 2011, Mediene et al., 2011). Despite the damaging limiting factors for this crop, biotic interactions have recieved little attention (Valantin-Morison and Meynard, 2008) and most of the studies on this crop underline abiotic factors such as nitrogen and water. Few extensive studies have investigated the effects of the whole crop management on WOSR in organic systems (Valantin-Morison and Meynard, 2008), accounting for the current lack of pesticide-free crop-protection strategies for this crop. Therefore, designing ecologically-

* Corresponding Author

sound and innovative pest management strategies, mobilizing several techniques, has become a major and urgent issue for organic Winter Oilseed Rape; these strategies should be adapted to diverse climatic and soil conditions.

Among the methods available for designing new crop management systems, model-based methods are often published (Loyce et al, 2002a, b, Chatelin et al., 2005, David et al., 2004, van Ittersum et al., 2003, Le Gal et al., 2010) and reviewed (Ould-sidi and Lescourret, 2011) demonstrating that crop models are useful tools to achieve this goal. Different crop models exist for Winter Oil seed rape (Gabrielle et al., 1998, Jeuffroy et al., 2006) but none of them takes into account the pests damage on the yield and the crop management impacts on pests; none of them has been built as a tool to help in designing free-pesticides crop management. Therefore, agronomic and new technical knowledge has been recently accumulated either by scientists, technical advicers and by farmers themselves. This expert knowledge could be used through the prototyping methodology (Vereijken, 1997), which combines regional diagnosis to identify the constraints and objectives, expert knowledge to build innovative cropping systems and on-farm experimentation to assess and adjust the system. This method can quickly produce innovative cropping systems, and disseminate them among pilot farms (Le Gal et al., 2011); it has successfully produced several innovative crop management systems or cropping systems (Dejoux et al, 2003, Lançon et al., 2007, Rapidel et al., 2009, Valantin-Morison, 2011). In our study case, previous research had produced experimental results on the impact of the date of sowing and nitrogen availability on diseases, slugs and weeds (Aubertot, 2004; Dejoux et al., 1999). Studies of the effect of cultivar on insects (Dosdall *et al.*, 1996, 1999; Barari *et al.*, 2005; Cook *et al.*, 2002, 2006) are also available. Nitrogen fertilisation has received much attention at the end of the 90's and it is now possible to adjust the date and the quantity of nitrogen thanks to several tools (Reau et Wagner 1998 ; Jeuffroy et al., 2006, Valantin-Morison et al., 2003, Rathke et al., 2006). Expert knowledge on mechanical weeding is also available (Lieven et al., 2008). Moreover, a regional diagnosis study in organic WOSR has pointed out the key factors that should be considered (Valantin-Morison and Meynard, 2008): competition for nitrogen due to weeds and nitrogen availability in the system. Despite this, the combination of all cultural practices in a crop management system has obviously received little attention for this crop, except the studies of Dejoux et al., 2003 and Valantin-Morison, 2011 for conventionnal farming.

This paper presents (1) a conceptual model of the effect of cropping techniques on WOSR, based on this knowledge;(2) two prototypes of crop management for organic WOSR designed using this conceptual model (3) the assessment of the agronomic performance of those two crop management systems in a farmers' field network.

## 2. Materials and methods

### 2.1 The conceptual model

To elaborate a new crop management system (CMS), scientists had to consider simultaneously cropping techniques, soil-climatic constraints, growers' objectives and their interactions. In this systemic approach of the agro-ecosystem, a conceptual bio-technic model is a useful representation to help in designing new crop management systems. This approach was adapted to organic WOSR, like it has been done for cotton or cacao Rapidel et al. 2006. In our study case, the conceptual model focus on the relationships between crop management, the environment, the crop and the different weeds, pests and diseases.

This conceptual bio-technic model has been designed thanks to the knowledge coming from the regional diagnosis on organic farming (Valantin-Morison and Meynard, 2008) and on conventional farming (Dejoux et al., 2003) and from the factorial experiments enlightening about the effect of different cropping techniques on diseases or weeds (Aubertot, 2004, Dejoux et al., 1999, Ferré et al, 2000). We have also combined the expertise of technical advicer on mechanical weeding (Cetiom, 2008, Lieven et al., 2008) and knowledge of the organic farmers of the network on their own soils, and on weeding techniques. On the contrary of Lancon et al., (2007) and Rapidel et al., 2009, we have not organised an "expert" session, gathering scientists, advisers and farmers, considering that the results of the diagnosis on farmers' fields allowed to design the frame of the conceptual model. Farmers' contribution to the conceptual model was restricted to the incorporation of their know-how on their pedo-climatic conditions and their reaction to the crop management systems proposed. Based on this conceptual model, two prototypes were designed. They were submitted to the farmers, improved according to their advice and tested on their fields; the first one, named "de-synchronization strategy", was based on the avoidance of weeds emergence, disease and slugs thanks to an early sowing, while the second one, named "mechanical control strategy", was based on the destruction of weeds and slugs before sowing and during the autumn thanks to false bed and hoeing. These two CMS aimed at increasing yield performance and reducing pests' impact.

## 2.2 The experimental design

The network of experiments was spread over 14 farmers' fields, in 12 farms (Figure 1), in 3 regions of France with contrasting climatic and soil conditions (fields M,M,N = Eure-et-Loir in the western Paris Basin; E,B = Puy-de-Dôme in the Center of France ; O,P,Q,R,S = Yonne in the south-eastern Paris Basin). It was studied over a two-year period (2001-2002 and 2003-2004).

As there was not any "reference organic crop management", there was no "control" crop management to compare to the innovative systems. Therefore the two prototypes of crop management systems (CMS) were compared to one another in each site. They were implemented without replicates, on two large plots of farmland (plots sized from 0,5 ha to 2 ha). The two plots were located on a homogeneous area of a same field. The management decisions were based on objectives and decision rules, as recommended by Meynard et al., 1996 and Debaeke et al., 2011. They differed on soil tillage before sowing, date of sowing, plant density, row spacing and mechanical weeding (table 1). Each experiment was named with a letter-figure code, in which the letter is a "field" code and the figure refers to the "harvest year" code (02, 04 for 2002 and 2004 respectively- figure 1). Even if the letter code of the field is the same between the two years (L02 and L04 or B02 and B04), WOSR was not implemented in the same field, because of crop rotation : this only means that fields belong to the same farmer. In 3 cases, farmers tested two soil tillage before CMS: shallow tillage (L02, B02, N02) and deep tillage (L02bis, N02bis, B02bis). Crop management plots, within each experiment, were designated with a little letter-figure code ((a) for the "desynchronisation" prototype and (b) for the "mechanical control" prototype– table 1). Crop management systems were implemented by the farmers while the description of each system were made by the team of researchers. Because of the difficulties to sow turnip rape as a trap crop (availability of the seeds and difficulties of work organisation), only three farmers have tested this landscaping arrangement : L02 and L04 in the west of Paris and B02, B04 in the center of France.

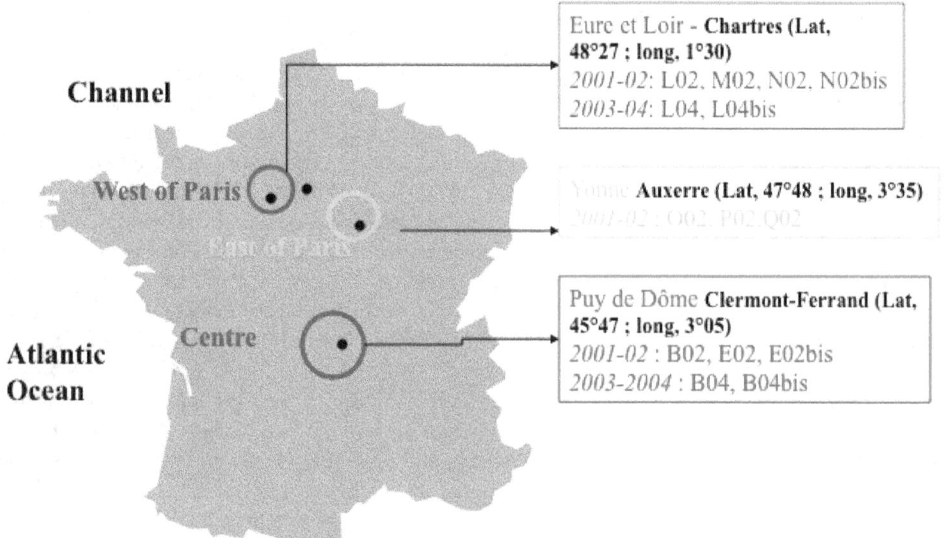


Fig. 1. Map of the fields experienced

### 2.3 Measurements and observations to assess the agronomic performances of the prototypes

The measured variables were selected to assess the agronomic and economic performances of the prototypes. In order to test whether the objectives assigned to each crop management systems (that are summed up in table 2) were achieved, agronomic measurements and pest damage were made on the crop and on the pests all along the crop cycle : biomass and nitrogen accumulated in the crop and in the weeds, weeds densities, phoma occurrence and severity, and insects' occurrence, yield and its components.

For each plot of CMS, in each farmer's field, whole plants were sampled at four different period of the year: (i) in early winter, before winter drainage and the occurrence of cold temperatures (from early November in the center of France to early December in the western region, which corresponds to a plant growth stage around 19 (Growth stage Lancashire et al., 1991), (ii) in late winter (from the end of January in the center region to mid-February in the western region - growth stage around 21), (iii) at early flowering (Growth stage 65 – beginning of April), and (iv) during pod filling (GS from 71 to 75 – beginning of June). In each field and for each CMS, at each period, samples were taken from six microplots of 0.5 m². The plants were counted and the roots separated from the aerial parts after washing. Dry biomass (after 48 hours of drying at 80°C) and the total nitrogen content of green aerial parts and of tap roots (Dumas method) were determined for each microplot sample. The numbers of branches, flowers and fertile pods per plant were counted at the pod-filling stage on 15 randomly chosen plants per microplot. The crop was harvested with a combined harvester, and, for each crop management system, yield was estimated from two samples of 150 $m^2$ to 400 $m^2$ for each CMS.

Weed infestations were assessed on the six microplots used for plant sampling, at the same dates. We identified the species, counted the individuals of each species and determined the total aerial dry biomass and total nitrogen content of all weeds. Shannon index has been calculated on weeds, which have been identifed until the species. Slugs have been counted on 3 traps per CMS early after crop emergence and later in autumn. Crop damages by pests and diseases were estimated on the sub-samples of 15 plants: we counted the number of plants in which larvae of root maggot, of cabbage stem flea beetle or of rape stem weevil were present. The occurrence of *Leptospheria maculans* (named also phoma or stem canker) in autumn was assessed by counting the number of plants with at least one leaf lesion. At pod filling stage, we counted the number of fertile pods, the number of podless stalks on 60 randomly sampled plants per micro-plot that gives information about the blooming flower abortion linked to the incidence of blossom pollen beetle damage (Nilson et al., 1994) and the number of burst pods in order to assess the pod midges damage.

## 2.4 Weather conditions for the network

For each field, the nearest meteorological station from either the Meteo France network or the INRA network was chosen. For each period of the crop cycle, the mean temperature and the numbers of days with frost or rainfall are indicated in table 2. The two years were rather similar : as an axample, for the west of Paris Basin, the number of rainy days reached respectiveley in 2002, 192 and in 2004, 176 days. The mean temperatures in the several period were similar between the two years for each regions but he center of France is the colder region in winter : the number of days with temperature below 0° reached 41 days in winter and 2 or 3 days at the beginning of the spring. In none of the regions a water deficiency occurred during winter, but in the Paris Basin, the water deficiencies were rather high during the autumn period specially the second year (table 2). Similarly for the spring period, every region displayed a water deficiency after flowering but with the maximum intensity in the center of France each year. The two years were rather similar in temperature for each region, but during the second year periods, dry weather were longer than in the first year, specially in the west of Paris.

## 2.5 Statistical analysis

The continuous data such as dry biomass of crop or weeds or nitrogen accumulated were normally distributed (Shapiro-Wilk's test) and the variance heterogeneity was constant (Bartlett test). Therefore we applied standard analysis of variance procedures. Proportions, such as the proportion of plants attacked by insects or diseases,were analysed after an arcsin√ transformation to stabilise variance between groups while the number of leaves counted were transformed with log. A mixed linear regression procedure was used to describe each system variable (crop biomass, nitrogen, weed density) depending on fixed factors (prototypes) and random factors (site and year). All analyses were done with "lme" procedure in R program.

| Crop management systems | Prototype (a) = "desynchronisation" prototype | Prototype (b) = "mechanical control" prototype |
|---|---|---|
| General principles | -crop vigor (high growth and nitrogen uptake) to enhance crop competitivity against weeds<br>-pest avoidance (slugs, stem canker, autumn insects)<br>- trapping spring insects with turnip rape border strips | -weed avoidance by shifting the period which both cultivated crop and weeds can emerge<br>-crop vigor before winter to reach 8 leaf stage and to reduce the severity of the insects attacks<br>- trapping spring insects with turnip rape border strips |
| Decision rules on sowing date and organic manure | -Early sowing date : *Before 10th of august*<br>-High amount of organic manure before sowing (objective= 100kg/ha of nitrogen available) | -Normal sowing date : *recommended sowing date for each region*<br>- Destruction of weeds before sowing thanks to false bed<br>- High amount of organic manure before sowing |
| Decision rules on plant density * | Sow with normal row spacing (17cm) to cover the soil<br>Sow at 80 seeds/m² (high plant density to cover the soil) | Sow with large row spacing (<35cm) to use the mechanical hoe<br>In order to reduce the competition between the plants, sow at 60 seeds/m² |
| Decision rules on weeding | -When crop reach 3 leaf stage, use the weed harrow | - When crop reach 5 leaf stage, use the weed hoe |
| Similar decision rules for the two CMS | *Objectives : ensure a good WOSR emergence*<br>Depending on soil type, soil moisture and the equipment available plough (deep work) or disk harrow (shallow tillage) before sowing<br>*Objectives : choice a cultivar in accordance with the whole CMS*<br>Choice of cultivar with the following characteristics: very resistant to Leptosperia maculans, late flowering, reduced susceptibility to stem elongation before winter (in order to reduce the frost risk) and late re-growth ability (Mendel or Pollen).<br>*Objectives : reduction of weed density during crop cycle.*<br>If densities of weeds > WOSR density, use mechanical weeding ; otherwise do no use any tools<br>Depending on soil moisture and tools available on the farm, use decision table for mechanical weeding recommended by technical advisors (Lieven et al., 2008 – www.cetiom.fr)<br>*Objectives : Trap insects*<br>Depending on the sowing tools and the field area, implement a border strips of forage winter turnip rape all the field around. | |

Table 1. description of the decision rules for each crop management system.

| Year | Region | Period | Mean temperature | Number of days below 0°C | Number of days with rain fall | Water deficiencies |
|---|---|---|---|---|---|---|
| 2002 | center of France | during autumn | 13.7 | 0 | 34 | 23 |
| | | during winter | 2.7 | 41 | 55 | 53 |
| | | regrowth-flowering stages | 6.4 | 2 | 9 | -38 |
| | | end of flowering-harvest | 11.9 | 0 | 46 | -143 |
| | west of Paris Bassin | during autumn | 16.3 | 0 | 50 | -41 |
| | | during winter | 4.1 | 8 | 50 | 80 |
| | | regrowth-flowering stages | 8.7 | 0 | 47 | 11 |
| | | end of flowering-harvest | 14.6 | 0 | 45 | -155 |
| | East of Paris Basin | during autumn | 16.1 | 0 | 43 | -10 |
| | | during winter | 4.1 | 27 | 44 | 46 |
| | | regrowth-flowering stages | 9.5 | 0 | 5 | -60 |
| | | end of flowering-harvest | 15.7 | 0 | 72 | -125 |
| 2004 | Center of France | during autumn | 13.7 | 0 | 45 | 47 |
| | | during winter | 2.7 | 41 | 77 | 232 |
| | | regrowth-flowering stages | 5.7 | 3 | 28 | 22 |
| | | end of flowering-harvest | 12.4 | 0 | 29 | -239 |
| | west of Paris Basin | during autumn | 15.6 | 0 | 29 | -122 |
| | | during winter | 5.7 | 7 | 72 | 346 |
| | | regrowth-flowering stages | 6.1 | 5 | 42 | 22 |
| | | end of flowering-harvest | 13.6 | 0 | 33 | -117 |

*Autumn was defined as the period from sowing to the first 5 successive days with Temperature below 0°C*
*Regrowth was defined as the first day of the 5 successive days with temperature above 0°C.*
*End of flowering was considered as the 30th of april*

Table 2. Climatic characteristics of the different regions during the period of experiment.

## 3. Results and discussion

### 3.1 The conceptual bio-technic model of a WOSR crop

We considered the crop and the different pests (weeds, diseases, slugs, insects) to be the biological system influenced by environment and by the technical system (crop management) (figure 2). This conceptual model was not designed to be exhaustive but it highligths the main interactions between the crop management, the plant population, the pests (weeds, insects, slugs and diseases) and the environment which helps in understanding the functionning of the "agro-ecosystem". It points out the main limiting factors, that has been identified in the previous study (Valantin-Morison and Meynard, 2008) and by farmers (insects such as cabbage stem flea beetles and pollen beetles). Therefore, the focus of the biological system is clearly on the plant-pest (insects or pathogens or weeds) interactions. It is composed of two compartments: the technical system and the biophysical system. The technical system is a combination of cropping techniques which act individually or in interaction with other ones (for example date of sowing and nitrogen supply interact on growth, competity against weeds ans stem canker development, see below). The biological system is influenced by climate (but not represented) and by technical system. In the biological system three compartments are distinguished : crop, pests and yield components, with each being represented by one or several variables.

Considering the previous study on organic WOSR by Valantin-Morison and Meynard (2008), the main limiting factors are weeds and nitrogen. The other limiting factors were slugs and 3 insect species : cabbage stem flea beetles (*Psylliodes chrysocephala*), rape stem weevil (*Ceutorhynchus napi*) and pollen beetle (*Meligethes aeneus*). But these limiting factors were rather rare in our fields' network (Valantin-Morison et al., 2007) and they were mainly mentionned by farmers and technical advisors. Depending on the type of pests, the damage could be significant on crop vigor (slugs) or on the number of the branches (cabbage stem flea beetles and rape stem weevils) or on the number of flowers (pollen beetles). We will describe the relationships between the technical system and the biological system for each limiting factor (figure 2).

In arable crops, weed demography and weed species are known to depend mainly on cropping history of the field and specially crop sequence and tillage interactions (Mediene et al., 2011 and Chauvel et al., 2001). Although few studies have dealt with the effect of soil management on weed emergence in WOSR, shallow tillage has been shown in corn and wheat crops to result in a larger weed seed bank than ploughing (Barberi and Lo Cascio, 2001; Feldman et al., 1997). Weed emergence flushes in crops can also be avoided or reduced by combining a delay of sowing to let the first germination occur before sowing (Chauvel et al., 2001) and a soil tillage before sowing to destroy the weed flushes. During the crop cycle, two categories of cultural mananagement operations could be used: (i) those reducing the number of weeds by chemical and physical control (see mechanical weeding in figure 2) or by shifting the period which both cultivated crop and weeds can emerge (see timing in crop cycle in figure 2) and (ii) those increasing the crop competitivity for light and nutrients (see crop vigor in figure 2). Numerous mechanical tools exist now to destroy the weeds; they work the inter-row space or inside the crop rows and are often efficient on young weed seedlings in dry soil conditions (Kurstjens ans Kropff, 2001). Mechanical weeding, because of its lower efficiency compared to herbicides, must be associated to other cultural operations. Maximising crop competitive ability can be maximised thanks to a combination of early sowing, dense sowing and high nitrogen availability during autumn. Winter oilseed rape is known to have a weed suppressive potential even in cover crop utilisation and specially when nitrogen is largely available in the soil (Kruidhof et al., 2008), thanks to its ability to catch nitrogen. Moreover, the date of sowing has been reviewed to have a major effect on plant growth at early stages (Dejoux et al., 2003) and the competitiveness of the crop (Whytock, 1993; Ferré et al., 2000). Nevertheless, there is an interaction between sowing date and soil nitrogen availability (Dejoux et al., 1999; Ferré et al., 2000). In cases of low levels of soil nitrogen, crops sown early suffered nitrogen deficiency earlier than crops sown on the normal date. This could have a direct effect on the aerial growth and competitiveness of the crop, opposite to the effect observed in cases of high soil nitrogen availability. Despite high sowing rate is rarely reported to increase the weed supressive ability (Dejoux, 1999, Kruidhof et al., 2008), plant density could have a significant effect on weed biomass and weed density especially when high weed density and large range of weed types are observed (Primot et al., 2005).

The effect of weeds on arable crop yields' is largely documented (Lutman et al., 2000, Vollmann et al., 2010, Vasilakoglou et al., 2010). In the study of Valantin-Morison and Meynard (2008), WOSR yields in organic system appeared to be strongly related to the variations of the number of flowers and the numbers of branches. Weed competition during autumn and nitrogen availability in soil during autumn were the principal factors accounting for the high variability of WOSR biomass at the beginning of winter and the number of branches.m$^{-2}$ at the end of winter.

Similarly to weed, soil-borne diseases occurrence and severity on major crops are often related to crop sequence and soil tillage, this combination playing a crucial role by affecting the sources of primary inoculum (Meynard et al., 2003). Considering cultural control during the crop cycle, crop management could reduce the crop vulnerability, that is in the case of disease the probability to be contaminated by primary inoculum. One of the strategies underlying pest avoidance is the desynchronization between crop susceptibility periods and the biological cycle of pests. In WOSR, the early sowing date permits to shift the periods of highest crop susceptibility to *Leptosphaeria maculans* (Aubertot, 2004). As the primary inoculum of *Leptospheria maculans* generally peaks between September and December (West et al. 2002), early crop sowing leads to a lower risk of infection just after emergence than other sowing dates. Another strategy to reduce crop vulnerability is based on the reduction of Leaf Area Index (LAI) of crop to cut down the number of leaf lesions, which will decrease the disease severity (Aubertot, 2004). Moreover indirect effects of nitrogen can also be foreseen to explain the general relationship (e.g. Agrios et al., 1997) between high N level and diseases: modifications of micro-climate due to high LAI; higher sensitivity to frost that can create wounds easing disease development. Endly, the effect of disease on arable crop yields is also well known and it is often related to a reduction of the grain mass.

Considering insects, the interaction between crop management, plants and pests is rather different since the ability of the pest to choose the host plant plays a crucial role. The cultural control is based on two strategies: (i) pest avoidance thanks to the desynchronization between crop susceptibility periods and the biological cycle of pests and (ii) pest repellent thanks to host plant quality manipulation. A complementary approach aims at enhancing the natural ennemies thanks to crop management and landscape management, but it is not concerned in this description.

Pest avoidance thanks to early sowing is well known: Dosdall and Stevenson (2005) demonstrated that the sowing date of oilseed rape strongly affects flea beetle (*Phyllotreta cruciferae*) damage. Indeed, the damage was greater on spring-sown oilseed rape than on plants sown in the autumn. Flea beetle feeding damage to oilseed rape apical meristems can prevent a compensatory response, but by the time of greatest injury, winter oilseed rape had well-developed, enlarged apical meristems making them less susceptible to damage. But sowing date effects can be antagonistic when considering different pest populations. In this way, Valantin-Morison et al. (2007) have shown that sowing oilseed rape early tended to increase root maggot (*Delia radicum*) damage, whereas it was associated with a lower level of attack by cabbage stem flea beetle. Dejoux et al., 2003 has also noticed that very early sowing crops were not destroyed by slugs on the contrary to normal sowing crops, less developped and so more vulnerable when the weather becomes favourable to slugs. The capacity of insects to identify a host plant suitable for its feeding and reproduction depends on the morphological and/or metabolic characteristics of the plant. The pollen beetle locates its host plant through visual and olfactory signals (Evans and Allen-Williams 1989, 1998). The beetles are principally attracted by the yellow colour of the flowers and by certain chemical signals released by the plant. It has been shown that degradation products of glucosinolates attract insects specialised on cruciferous host plants (Feeny et al. 1970, Finch 1978, Free and Williams 1978). Based on the hypothesis that the production of glucosinolates by cultivars of winter oilseed rape (WOSR) and other Brassicaceae may attract pollen beetles, many studies have focused on the effects of host plants on insect orientation and feeding (Bartlet et al. 2004), oviposition behaviour (Borg and Ekbom 1996), and egg production of the pollen beetle (Hopkins and Ekbom 1999). Turnip rape (*Brassica rapa*) has been found to attract more pollen beetles in both laboratory and field conditions (Hokkanen 1989, Cook et al. 2002, 2006, Valantin-Morison and Quéré, Rusch and Valantin-Morison 2009). The same effect has been reported for other oilseed rape pests such as cabbage stem flea beetle (*Psylliodes chrysocephala*) (Büchi 1995, Barari et al. 2005) and cabbage seedpod weevil (Carcamo et al. 2007). Turnip rape is thus often used in this particular situation as a so-called trap crop. Simulations using a spatially explicit individual-based model show that for herbivores that actively immigrate from a nearby source via the field edge, a surrounding border trap crop is the optimal arrangement (Potting et al. 2005). For more details one can refer to the review on biological control in Oilseed rape (Rusch et al. 2010a). Endly, the impact of insects on yield is not well documented for WOSR except for pollen beetles (Hansen et al., 2004) but we could propose several hypothesis, related to the impact of each pest on the yield components: cabbage stem flea beetles reduce the crop biomass in autumn and the number of branches during re-growth stages, pollen bettles reduce the number of flowers, stem weevils reduce the number of branches, resulting in the decrease of the number of grains.

One can notice that the interactions – between cultural operations, or between cultural operation and crop status – are important and taken into account in this scheme (ie interaction between sowing date and nitrogen supply on weed suppressive ability of the crop). In addition, one technique can be advantageous to limit one category of pest populations (early sowing and cabbage stem flea beetle), but can be detrimental to the control of other (early sowing and root maggot). Instead, many of those interactions and antagonistic effects must be taken into account when designing pest management strategies thanks to this type of conceptual framework of integrated crop management.

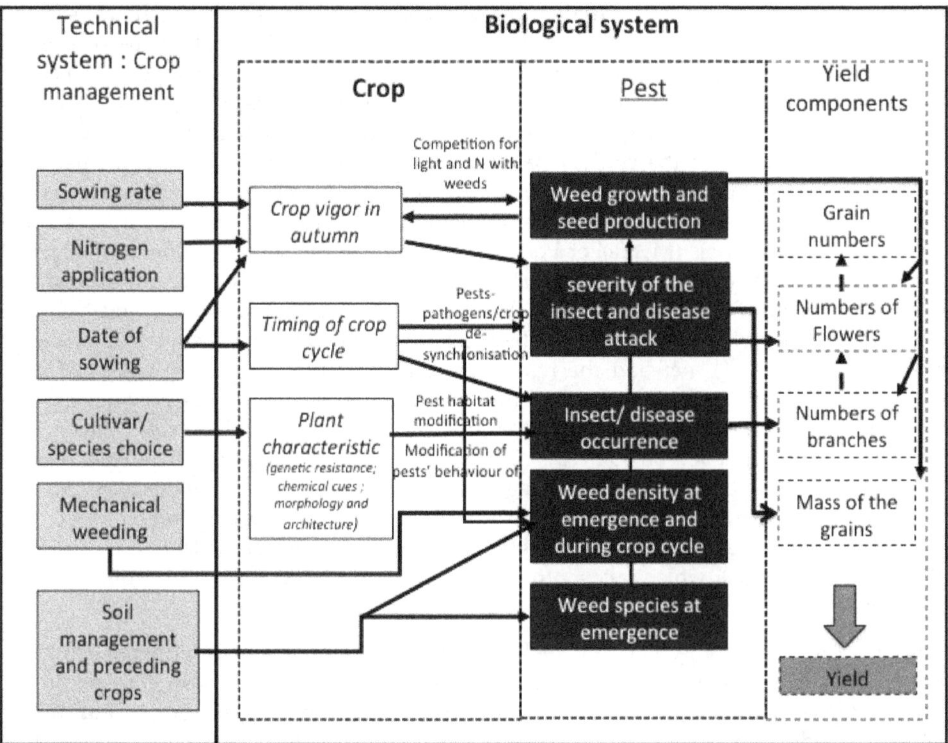


Fig. 2. Schematic representation of the Conceptual bio-technic model of the relationships between crop management, plant and pests. Mecanisms which related the interaction between technical system and crop / pests system are explained on the arrow.

### 3.2 Description of the crop management systems

In accordance to this conceptual model, two theoretical prototypes have been designed and tested on farmers' fields. They were built in order to limit the impacts of the main limiting factors, i.e nitrogen deficiencies, weeds and insects, mentionned before, in order to maximize yields and gross margins. Following the protoyping method, the cultural operations were combined to design theoritical crop management system (CMS). The prototypes were described either by the principles linked to the combination of cultural operations, by the order of implementaion of the cultural operations and by decision rules for some specific and single cultural operation.

Prototype (a), named "desynchronisation strategy", was built on three principles. (i) The first one was to enhance crop vigor during autumn, (ie crop biomass and LAI) in order to enhance the weed suppressive ability of the crop and to reduce the severity of the insects and slugs attacks; (ii) the second one was to desynchronize pests attacks (slugs, cabbage stem flea beetle and *Leptospheria maculans*) occurrence and susceptible stages of the crop. (iii) Endly, since pest avoidance might be efficient only on autumn insects, we aim at trapping spring insects. Suitable combination of individual cultural practices were deduced from the relationships between the crop variables in the biological system and the technical system. The crop management system was based on a combination of early (before 10th of august) and dense sowing, with organic manure or effluent before sowing. Turnip rape border strips were sown to trap spring insects, and a late flowering cultivar oilseed rape,was chosen to maximise the differences of stages between the border strips and the center of the field). The specific characteristics of the cultivar resulted in a very narrow choice of cultivars: very resistant to Leptosperia maculans, late flowering, reduced susceptibility to stem elongation before winter (in order to reduce the frost risk) and late re-growth ability; we used the two same cultivars in all trials (Mendel or Pollen, depending on the region and on the year). Soil tillage was reduced before sowing, linked to the short period between the cereal harvest and early rape sowing. But we were unable, based on our scientific knowledge and technical know-how, to formulate rules that could be applied nationwide. In each trial, decisions were taken jointly by the farmer and the research team, so as to ensure a high level of crop emergence. Generally, the soil was tilled immediately after harvest of the previous crop to bury stubble and straw and to allow the emergence of cereal volunteers. The type of implement used depended on soil type, soil moisture and the equipment available: plough (deep work) or disk harrow (shallow tillage). Shallow tillage is more often used in the first year than in the second one. Endly, concerning the mechanical weeding (type of tools and date for weeding) we used a decision table formalised by technical advisors (Lieven et al., 2008 ; www.cetiom.fr), which takes into account crop and weed stages, and soil moisture. The CMS (a) is detailled in table 2.

Prototype (b), named "mechanical control strategy", was built on three principles: (ii) The first one was to shift the period which both cultivated crop and weeds can emerge; (ii) the second one was to reach 8 leaf stage before winter and to have enough crop vigor during autumn to reduce the severity of the insects attacks; (iii) the third one was to trap spring insects. The sowing date of CMS (b) was a normal one, i.e. about 20 days later than CMS (a), with organic manure before sowing. The soil was tilled immediately after harvest of the previous crop to bury stubble and straw and to allow the emergence of cereal volunteers (plough or disk harrow). After this initial tillage the soil was turned by numerous shallow tillage operations (disk harrow or cultivator) in order to destroy emerging weeds, The crop was sown at a medium density, with large rows, to permit the weeding harrow. As in prototype (a), a late flowering cultivar was associated with turnip rape border strips. The cultivar choice was based on the same characteristics as CMS (a) and the same Pollen and Mendel cultivar were implemented, depending on the region and the year.

The success of these 2 prototypes underlies on three main hypothesis that we have examined with data collected in farmers' fields:

**Hypothesis 1**: Prototype (a) allows a better and quicker growth of rapeseed in autumn which permits to cover the soil and smoother weeds, while prototype (b) will reduce the weed population by its destruction before sowing. The efficiency of the two prototypes for weed management are supposed to be similar. The way to assess this hypothesis is to measure the weed biomass and densities in autumn and at the beginning of spring.

**Hypothesis 2**: Prototype (a) is supposed to reduce better autumn pests, disease incidence and WOSR vulnerability thanks to pest avoidance and crop vigor. The way to assess this hypothesis is to record the number of leaf lesion of *Leptospheria maculans* and the occurrence of insects damage in autumn.

**Hypothesis 3**: This hypothesis concerns the landscaping of the field around. Turnip rape trap crop around fields allows to reduce blossom pollen beetle damage better than other elements of crop management. The way to assess this hypothesis is to record the number of pollen beetle in trap crop and in WOSR.

### 3.3 Assessment of the crop management systems

#### 3.3.1 Evaluation of the performance of the systems: Yield and its components

Organic WOSR yields were very variable and sometimes very low compared to conventionnal system (Fig. 3a): It ranges from 0.1 to 3.4 t/ha with a mean of 1.3 t/ha ± 0.95. In conventionnal systems, the mean yield of WOSR in France ranged from 2.9 to 3.5 t/ha during those previous 10 years (Ecophyto R&D, 2009). However, the organic yield of such difficult crop is similar to these results in France (Reference of national agency of organic farming, www.agencebio.fr) and in Suitzerland (Daniel et al., 2011). The differences between the two CMS are rather low and not significant (mean of CMS (a) : 1.4 t/ha ± 0.9 mean of CMS (b) = 1.3t/ha ± 1.1). The highly significant relationship between the number of grains per square meter and the yield (fig. 3b) suggests that the number of grains is the main determinant accounting for yield variations, which could means that limiting factors of the yield resulted in a decrease of the grain numbers rather than an impact of grain weigth. In figure 3a, one can observed that field B (in the center of France) exhibited systematically a yield in CMS (a) lower than in CMS (b) whatever the year, while it is the opposite for the field L (in the west of France). This result will be examined thereafter in the light of weed pressure. It is also noticible that three fields have not been harvested, mainly because of invasion of weeds (L02-CMS(a)), slugs (O02-CMS(b)) and pollen beetle damages (N02).

Considering that the number of grains is determined by several yield components, (the number of branches/m$^2$, the number of flower/m$^2$ and the number of pods per square meter), we compared the two CMS in the establishment of those components (Fig. 4). In this figure, the two CMS seem to be similar and the linear mixed model did not pointed out any significant CMS effect, only a year-site effect. However, it is interesting to examine several field results. The very low yield recorded on fields E02 and P02 could be related very early in the cycle to a reduction of the number of branches/m$^2$ (fig. 4a) and the number of flowers/m$^2$ (Fig 4b) and thereafter to the number of pods per square meter (Fig 4c). More surprinsingly, the filed Q02 exhibited the lowest yield while the number of branches and the number of flowers displayed respectively high values (from 100 to 400 branches/m$^2$) and medium values (3000 to 9000 flowers/m$^2$). Significantly the number of pods per square meter were particularly low for this field, which is a good explanation for the poor yield performance of field Q02. The high proportion of burst pods due to pod midge attack (from 16% to 23% on this field while it reached hardly 10% on the other fields) might be responsible for such reduction of pod/m$^2$. The highest yield recorded on fields L04 and L04bis are also in accordance with the results on the successive yield components, while the fields B04 and B04bis reached high values of yield and of number of branches/m$^2$ and relatively medium values of number of flowers/m$^2$ and pod/m$^2$.

#### 3.3.2 Crop growth and competitivity against weeds

The weed biomass in early winter is shown in figure 5a. The number of trials for weed biomass is not similar on either side of the bissectrice line, the prototype (a- desynchronisation strategy)

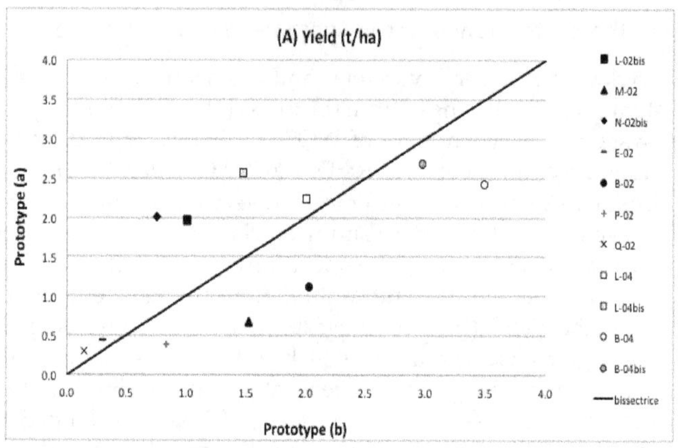


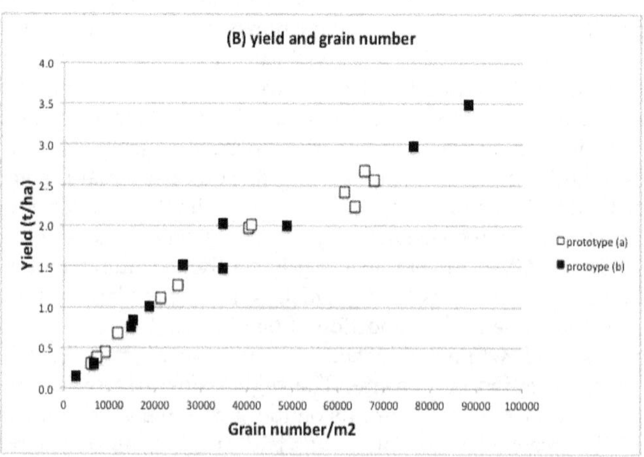


Fig. 3. Yield obtained for the two CMS in each field and relationship between grain number and yield (Two CMS has been destroyed: prototype (b) field O02, prototype (a) field N02 and L02.)

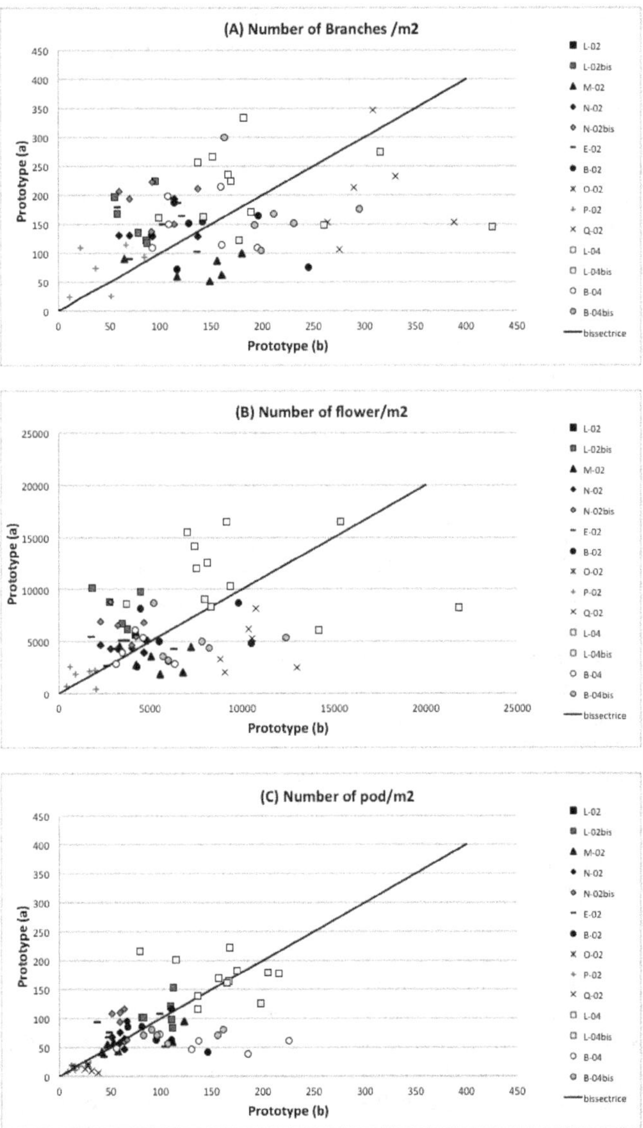


Fig. 4. Yield components of the two prototypes: (A) Comparison of the number of branches/m² in spring (B) Comparison of the number of flower/m²; (C) Comparison of the number of pods/m²

displaying higher weed biomass than the prototype (b- mechanical control strategy) and the first year exhibiting almost systematically higher weed biomass for the prototype (a- desynchronisation). The mean value of weed biomass for the prototype (a) reached 0,57 ± 0,42 t/ha while for the prototype (b) it did not exceed 0,36 ± 0,38 t/ha. Despite weed biomass exhibited a very high variation (figure 5a), the linear mixed analysis demonstrated a significant prototype effect (P<0,001) and a significant year-site effect (P<0,001). In the conceptual model, the weed suppressive effect which could induce a decreasing of weed biomass was assumed to be related to a crop vigor with high crop biomass, and an emergence of the crop earlier than that of the weeds (especially the broad-leaves weeds which emerge in autumn). The crop biomass, the nitrogen accumulation and the number of leaves produced since the emergence are shown respectively in figure 6a, 6e, and 6c. Early sowing exhibited higher plant biomass in numerous cases (means for the two years : (a)=3.07 ± 0.31 t/ha ; (b)= 1.84 ± 0.18 t/ha), which is in accordance with Dejoux et al., 2003), and is consistent with the studies of Ferré *et al.* (2000), Jornsgard *et al.* (1996), Van Delden *et al* (2002). However, this high biomass of the crop and the high nitrogen accumulation in the crop (Fig 6e) for the CMS (a) did not always resulted in a high suppressive weed ability. This suggests that the objective of the first CMS (a), that is enhance crop vigor to smoother weeds, did not succed in every field : sometimes, this strategy was less efficient than that of CMS(b), that is avoid weeds thanks to false bed and mechanical weeding. A covariance analysis (data not shown) demonstrated that soil management, weeding and nitrogen in soil at sowing could explain this variability in weed biomass. Table 3 summarizes these factors that could explain for this differences. According to the data of crop and weed biomass and soil nitrogen and plant density, four situations have been distinguished in table 3:

- For field trials N02, N02bis, L02bis, L04bis, L04, weed biomass was much lower on prototype (a) than on protoype (b). For those trials, crop exhibited very high values of crop biomass in CMS(a); plant density were systematically higher in CMS (a) than in CMS (b), the soil depth was high and and ploughing tillage has been used for the two CMS. Moreover, the high amount of nitrogen in soil was particularly noticeable for those trials(from 135 to 241 kg/ha).
- For fields Q02 and P02, crop biomass in CMS(a) was lower than in CMS(b) but weed biomass remained low for the 2 CMS. The weed densities were rather high (for instance 33 plants/m$^2$ and 37 plants/m$^2$ respectively for field P02, CM(a) and CM(b)), but the flora was not very harmful (main weed species recorded in the fields: lady's mantle, veronica, sow-thistle).
- For fields L02 and M02, despite a high value of nitrogen in soil, CM(a) did not resulted in a higher crop biomass than in CM(b), mainly because of shallow tillage, without false bed, before early sowing leading to the emergence of an important population of volunteers at the same period of WOSR emergence. On these fields, a long period without rainfall in august leaded to a very heterogeneous emergence of the crop. As a consequence, CMS(b) was characterised by a higher crop vigor than the CMS(a), and a lower population of cereal volunteers, destroyed before sowing.
- For fields, E02, O02, B02, B04 and B04bis, despite high values of nitrogen in soil, high crop biomass and nitrogen uptake, despite the use of mechanical weeding, both CMS(a) and CMS(b) exhibited a particularly high weed density (for examples from 80 to 140 plants /m2 in field B02) and the weeds species were presumably harmful: sanves, cereals volunteers, foxtail. Therefore we suggest that above a threshold of weed pressure and with very competitive weeds, neither a competitive WOSR crop nor mechanical weeding could permit to suppress weeds.

Despite these observations on the species, considering the diversity of weeds between the two prototypes, shannon index was not significantly different and ranged from 0 to 3,35 with a mean reaching 1,35.

All those results are consistent with the results in conventionnal farming, obtained by Dejoux et al., 2003 but demonstrate that the interactions between techniques need to be understood to propose innovative CMS. It appears that the efficacy of CMS (a) is clearly related to interactions between pedo-climatic environment of the crop and several individual cultural operations such as : date of sowing, nitrogen supplies and soil tillage. The strategy which consists in enhancing crop vigor during autumn, (ie crop biomass and LAI) in order to enhance the weed suppressive ability of the crop is possible thanks to early, dense sowing in specific conditions: when nitrogen supply is above 150kg/ha, in soil with high depth, when crop density is homogeneous and relatively high and if previous soil tillage has permitted to destroy volunteers of cereals or other harmfull weeds (deep soil tillage is recommended). When soil depth is lower than 60 cm and when nitrogen supply in soil at sowing is lower than 100kg/ha, the second strategy, which consists in destroying the weeds (i) before sowing and (ii) by mechanical weeding hoe after the crop emergence, might be more efficient.

| **description of weed and crop biomass** | **description of competition relationships** | **name of the field** | **plant density** | **Nitrogen in soil at emergence** |
|---|---|---|---|---|
| Weed biomass (a)<(b)<br>crop biomass (a)>(b) | crop vigor=>weed smoothering for prototype (a) | N02<br>L02bis<br>N02bis<br>L04<br>L04bis | More plants in CM(a) than in CM(b) | more than 150kg/ha |
| Weed biomass (a)<(b)<br>crop biomass (a)<(b) | low crop vigor but few weeds in (a) prototype | Q02<br>P02 | More plants in CM(a) than in CM(b) | more than 250 kg/ha |
| Weed biomass (a)>(b)<br>crop biomass (a)<(b) | crop vigor=>weed smoothering for protoype (b) | L02<br>M02 | less plants in CM(a) than in CM(b) | more than 150kg/ha |
| Weed biomass (a)>(b)<br>crop biomass (a)>(b) | crop vigor but no weed smoothering for prototype (a) | E02<br>O02<br>B02<br>B04<br>B04bis | very irregular | from 60 to 300 kg/ha |

Table 3. Diagnosis analysis of the differences between the two CMS in the weed management

### 3.3.3 Pest avoidance and occurence of pest damage

None of the plots sown early were destroyed by slugs while two plots sown normally after 1rst September were destroyed. Slugs counted at the beginning of October never exceeded 5 slugs/$m^2$ but there were 0.34 slugs/$m^2$ on early sowing plots while there were 1.3 slugs /$m^2$ on normal sowing ones. The pest avoidance objective of the CM(a) has been achieved. It confirmed the suggestions, made by Dejoux *et al.* (2003) on very early sowing experiments in conventionnal farming, that were based on the frequencies of molluscicides treatments.

Concerning the cabbage stem flea beetle (*Psylliodes chrysocephala*), figure 5b shows that CM(a) was less attacked by cabbage stem flea beetle (fig 5b- mean for the two years : (a)= 0.13 ± 0.03 Confidence Interval 10%- (b) = 0.20 ± 0.04 Confidence Interval 10%). These results are consistent with findings of Dosdall et al., 1999 and Valantin-Morison et al., 2007 and could largely be explained by insects phenology. Adult beetles invade the crop in mid-September and October and attack newly emerged seedlings by chewing pits in the cotyledons, leaves and stems. Females lay their eggs in the soil from late August onwards and the emerging larvae bore into the petioles and mine the leaves and stems (Alford et al., 2003). For crops sown early in August, the chewing of the leaves was harmless, as the larvae were found in the petioles of leaves destined to fall earlier, before winter. Therefore in this case, the objective of avoidance of the pest thanks to early sowing proposed in the CM(a) has been achieved.

Concerning the root maggot (*Delia radicum*) (Fig 5c), the linear mixed analysis points out significant effects of CMS and of year-site either for the proportion plants attacked by the root maggot: CM(a) was more attacked by than CM(b). Moreover, we observed a very significant effect of CMS in the collar diameter and biomass in tap root: CMS(a) exhibited larger collar diameter (Fig 6b) and more biomass in roots (fig 6d). As a consequence we could argue that root maggot attacks occurred more frequently in CMS (a) than in CMS (b) but were less prejudicial for growth plants, thanks to a high root vigor. Those results are in accordance to those of Dosdall et al (1996) and Valantin -Morison et al., 2007. Dosdall et al. (1996) also showed that root maggot infestations were reduced by sowing Canola in late May rather than mid May and by high sowing density. This result could be explained either by insect phenology (Alford et al., 2003) and host plant preference: this insect lays its eggs in plots sown early and females prefer plants with large collar diameters. Therefore in this case, the objective of avoidance of the pest thanks to early sowing proposed has not succeeded but the plant vigor, observed in the CM(a) permitted to reduce the damage of the attack (figure 6c and 6d).

The number of plants with at least one leaf lesion of *Leptospheria maculans* were higher on normal sowing (Fig. 5d; 0.31 + 0.08 Confidence Interval 10%) than on early sowing (0.12 ± 0.08 Confidence Interval 10%). This result was largely related to the significant difference in the number of leaves produced between the CMS (Fig 6a): plants in CMS(a) produced more leaves than plants in CMS(b) and therefore were less susceptible to receive the first peak of spores during autumn.

As a conclusion, the hypothesis which argued that early sowing could allow pest avoidance was confirmed for cabbage stem beetle and stem canker. The hypothesis argued also that early sowing allowed higher growth of plants, which could be less prejudicial for plants could be confirmed for root maggot since the collar diameter of plants in prototype (b) was higher and that severe damage of root maggot was not observed even if they were more often attacked.

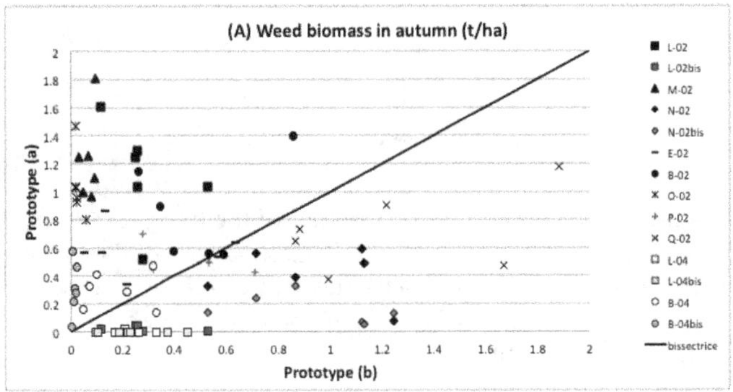


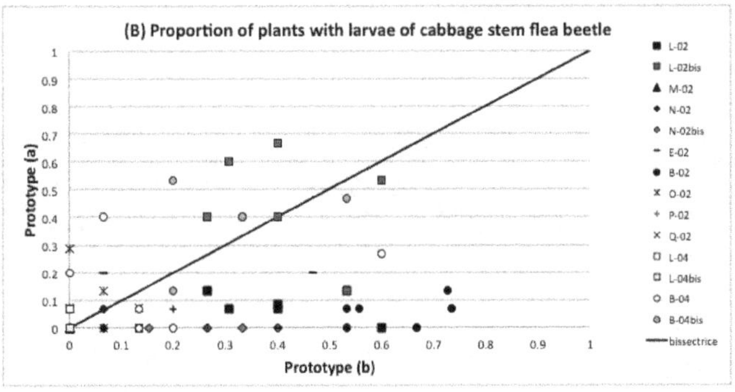


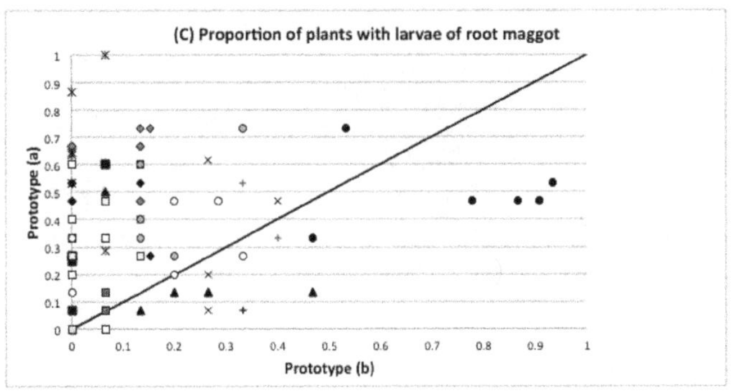

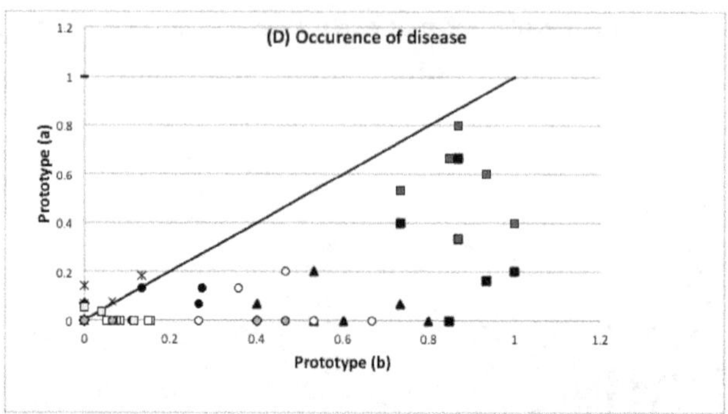


Fig. 5. Weed competition, Disease and pest occurence of the two prototypes : (A) Comparison of weed biomass in autumn (t/ha) (B) Comparison of the proportion of plants with at least one larvae of cabbage stem flea beetle; (C) Comparison of the proportion of plants with one larvae of root maggot (D) Comparison of the proportion of plants with at least one spot of Leptospheria maculans

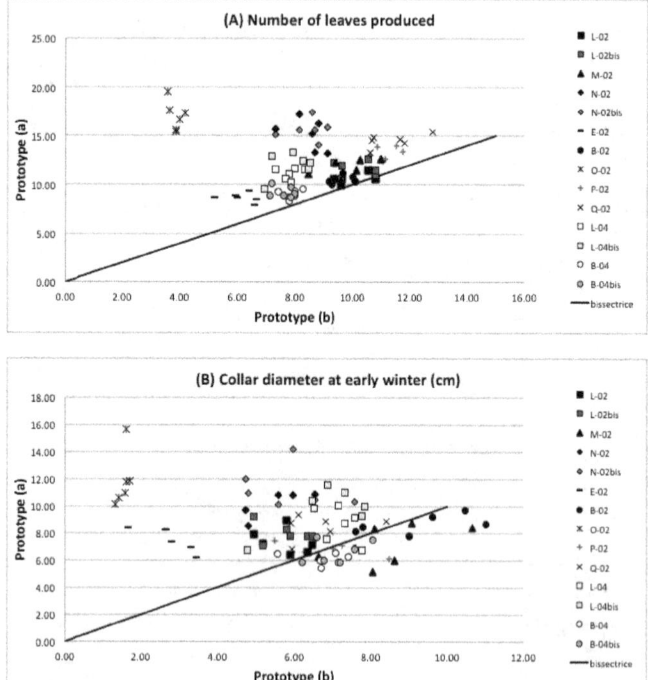

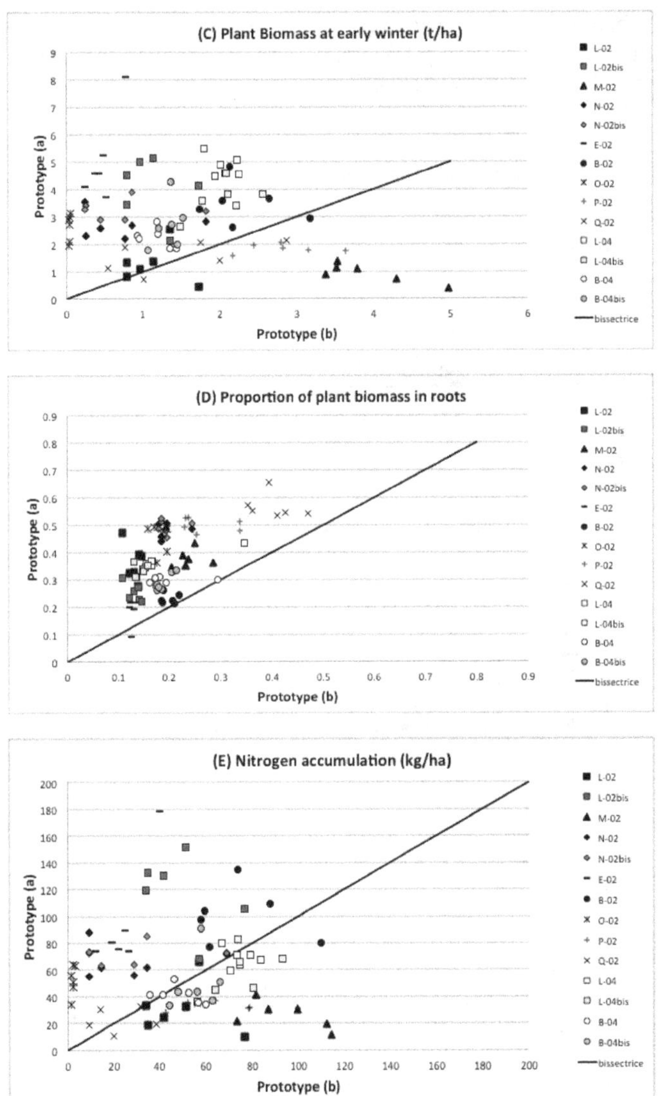


Fig. 6. Plant growth and development of the two prototypes: (A) Comparison of the number of leaves per plant in autumn (t/ha) (B) Comparison of the collar diameter; (C) Comparison of the crop biomass (D) Comparison of the proportion of biomass in tap roots. (E) nitrogen accumulation at early winter

### 3.3.4 Turnip rape as a trap crop to reduce the damage of insects in WOSR

Turnip rape border strips were sown to trap spring insects, in four fields. The fig 7a shows a significant effect of turnip rape on pollen beetle numbers in 2/4 fields, on cabbage stem flea beetle numbers in 2/3 (Fig 7b) and no significant effect for rape stem weevil damage (fig 7c).

The number of cabbage stem flea beetles and pollen beetles was higher on turnip rape, while rape stem weevils (*Ceutorhynchus napi)* damage seems to be more important on WOSR than on turnip rape, despite the no significant effect.

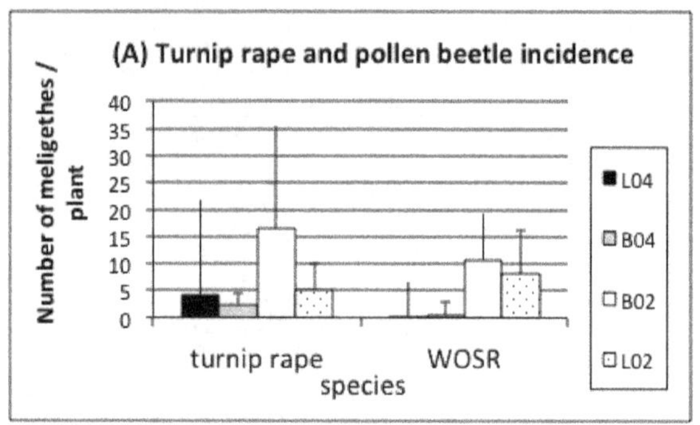


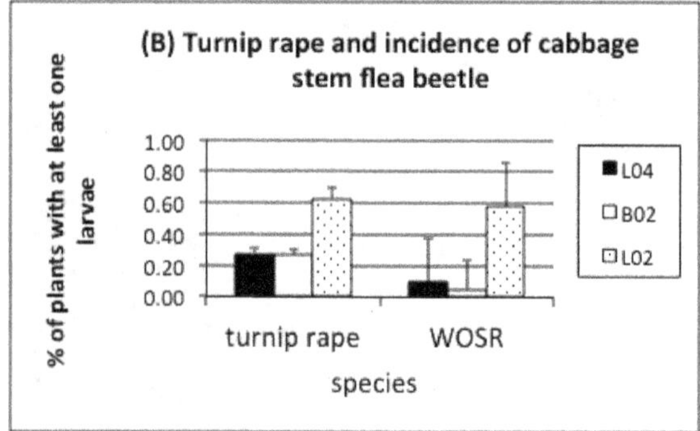


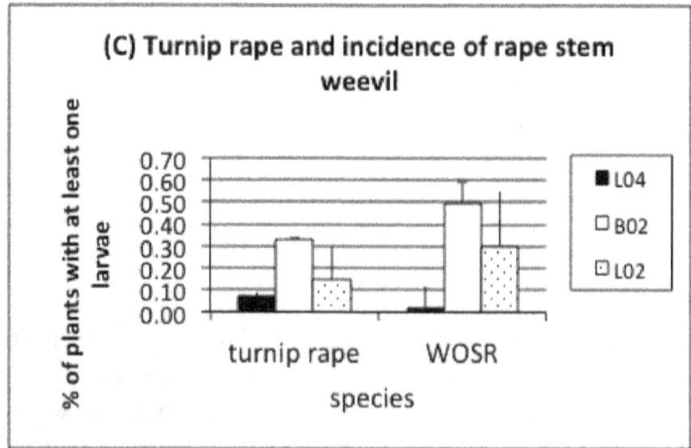


Fig. 7. Turnip rape effect on (A) pollen beetles number (*Meligethes aeneus*), (B) cabbage stem flea beetle (*Psylliodes chrysocephala*) (C) Rape stem weevil (*Ceutorhynchus napi*)

The pollen beetle, the cabbage stem flea beetles and the stem weevils are the most numerous of a suite of pests that attack oilseed rape and has been reported by the farmers and advisors to be the most damaging insects for this crop. For the past decade, control of those pests in France has relied exclusively on the pyrethroid class of insecticides, which is impossible for organic farming systems. Cook et al., (2009) points out that ecological approaches to control of pollen beetles emerging from such research include development of monitoring traps (so that growers can more easily and accurately identify when spray thresholds have been breached), trap cropping tactics, and practices that promote conservation biocontrol.

Trap crops are plant stands that are designed to attract, intercept and retain insects, thereby reducing pest density and damage to the main crop (Cook et al., 2007a). Knowledge of host plant location, acceptance and preferences of pests as well as their natural enemies is exploited in this tactic and has been developped in several studies in the 1990s indicated that turnip rape (*Brassica rapa*) was a preferred host plant of pollen beetles (Hokkanen et al., 1986; Hokkanen, 1991; Büchi, 1995). These field studies using a WOSR model indicated that turnip rape (*Brassica rapa*) shows good potential as a trap crop for pollen beetles as turnip rape plots were more heavily infested (Cook et al., 2006).

The results observed in this field network are in accordance with studies of Cook et al., 2006 and Barari et al., 2005. However, the none effect of turni p rape on rape stem weevils has never been reported. However, this trap cropping tactic suffered from several implementation difficulties: where and how many area of turnip rape could be efficient enough to control those pests ? What about the consequences at larger scales in the lanscape and for the future generation of pests ? Actually, Cook et al., highlighted that recent research developments have now led to a greater understanding of how the strategy works and will enable the development of robust and effective commercial-scale trap cropping tactics for both winter and spring oilseed rape production. An individual-based model developed at Rothamsted predicted that a border trap crop would be the most efficient arrangement of the trap crop, particularly since the pest infests the crop from the edges (Potting et al., 2005). Future studies will test the trap cropping tactic on a whole field scale in commercial systems and examine the effect of field size on the success of the method.

## 4. Conclusion

It appears that the utilisation of the conceptual model on the interactions pest-crop management and environment helped in building and discussing CMS with organic farmers. The agronomic performances of the CMS tends to be improved despite some low yield performances (P02, Q02, E02). The hypothesis which argue that early sowing could allow pest avoidance was confirmed for several pests: slugs, cabbage stem flea beetle and stem canker. The hypothesis argued also that early sowing allowed higher growth of plants, which could be less prejudicial for plants could be confirmed for root maggot since the collar diameter of plants in prototype (b) was higher and that severe damage of root maggot was not observed even if they were more often attacked. However, the ability of competitiveness agianst weeds for CMS (a) is clearly related to interactions between pedo-climatic environment of the crop and cultural operations such as : date of sowing, nitrogen supplies and soil tillage.

After this study, in the light of these results and considering the farmers' constraints either on soil type, on the availability of organic manure and their nitrogen contents and on labour organisation, on the availability of mechanical weeding tools, we proposed to the farmers to implement only, one type of CMS among the two (table 4).

| Soil depth | Availability of organic manure | Availability of mechanical weeding tools | Choice of CMS |
|---|---|---|---|
| Deep (>90cm) | yes | none | Ploughing before CMS(a) |
| Deep (>90cm) | no | yes | Ploughing before CMS(a) with large row spacing to use weeding hoe |
| Deep (>90cm) | no | none | Implement WOSR after legume to reach high nitrogen supplies during autumn + CMS(a) |
| Low (<60cm) | yes | yes | CMS(b) |
| Low (<60cm) | no | yes | Implement WOSR after legume to reach high nitrogen supplies during autumn + CMS(b) |

Table 4. Criteria at farm level and characteristics of soil type to choose the crop management system.

Concerning the insects control, the objective should be to optimise the natural pest control of the CMS and the resilience ability in order to manage such populations. Gurr et al., 2003 proposed to use two complementary ways to enhance natural pest control: "bottom-up" and "top down" approaches. The bottom-up effect of resource quality has been hypothesized as one of the factors determining herbivore abundance and population dynamics (De Bruyn et al., 2002). Because crop management influences resource quality through modification of local conditions, farming practices are assumed to affect pest density in the field (Rusch et al., 2010a,b). Particularly, the plant vigor hypothesis predicts that herbivores will be more abundant on the most vigorous plants (Price, 1991). For instance, pollen beetle have been found to prefer larger buds and plants at later development stages for feeding and ovipositing and to choose their host according to their glucosinolate content (Nilsson, 1994, Hopkins and Ekbom, 1999, Cook et al., 2006), suggesting host-plant selection according to specific quality requirements. In this study it has been proved that crop management variables such as sowing date, nitrogen nutrition and plant density influenced plant development (number of leaves, collar diameter, plant biomass), and thereafter influenced populations and damage at the field scale through host-plant quality. This indirect effect has also been highlighted in a previous study (Valantin-Morison et al., 2007). The « top-down » approach relies on the enhancement of natural pest regulation by natural ennemies. Enhancement of the natural regulation functions of agroecosystems appears to be one of the main ways in which we can decrease the use (Wilby and Thomas, 2002) of chemical pesticides for pest control and increase the sustainability of crop production. However, the factors responsible for the maintenance or enhancement of natural pest control remain unclear. Moreover, the environmental and economic benefits to farmers of increasing the activity of natural enemies of crop pests remains a matter of debate, in the absence of clear scientific evidence. It has been shown that community structure, species richness and abundance, population dynamics and interactions within and between trophic levels are affected by spatial context (e.g., patch size, spatial configuration, landscape composition, habitat connectivity or even the structural complexity of habitats) (Bianchi et al., 2006;

Tscharntke et al., 2007). There is a growing body of evidence that landscapes with high proportions of semi-natural habitats tend to support lower populations of pests than simple landscapes do, due to higher top-down control by natural ennemies (Bianchi et al., 2006) and Thies et al., 2003 has shown this on pollen beetles. In this study we did not focus on the field environment on pest populations, except through the implementation of turnip rape trap crop. But in a previous study it has been found that fields with woodland around exhibited more pollen beetles damage and it has been confirmed and explained in Rusch et al., 2011. This latter study alongside those from the literature suggest that host-plant selection by pollen beetle operates at very fine scales (i.e. selection between potential host plants within the field), whereas host-patch selection is mainly determined by the landscape context. In the ligth of those elements, we suggest that bottom-up effects, related to CMS at field scale, is involved in the ability of the crop to support, to reduce the damage of an abundance of pests determinated by the predominant influence of the landscape context. Thus, in an integrated pest management perspective both crop management and landscape context have to be taken into account together in order to identify and rank the relative importance of local and landscape predictors and their relative scales on pest population dynamics.

Finally, the agronomic performances assessment of such Crop Management system has to be completed by an economic and an environmental assessment like it has been done by Simon et al., 2011, for orchards and annuals crop by Debaeke et al. 2009, Aubry et al. 1998 and Munier-Jolain et al. 2008. The authors have generally used the comparison of organic, IPM and/or conventional systems to assess the performances and the environmental effects of crop management regimes. In such experimental system approaches, the gross margin, the economic efficiency and the pesticides use or impact, the level of inputs and farmers' labour has been quantified within fluctuating regulatory and climatic contexts.

## 5. Acknowledgment

We would like to thank G. Grandeau, V. Tanneau, B. Fouillen, C. Souin for technical assistance. Y. Ballanger from Cetiom (Centre technique des ole ́agineux métropolitains) provided helpful expertise for the diagnosis of insect damage. Financial support for this study was provided by INRA (CIAB) and CETIOM. We would especially like to thank R. Reau, L. Quéré , C. Bonnemort, and D. Chollet, from Cetiom for their help in finding financial support and in setting up the network of farmers. Endly, these experiments would not be possible without the organic farmers themselves, who were either very openminded and ready to suggest adaptation of the theoritical prototypes.

## 6. References

Alford, D.V., Nilsson C., Ulber B. (2003). Insect pests of oilseed rape crops. In D. V. Alford (Ed.), Biocontrol of oilseed rape pests (pp. 9–41). Blackwell Science, Oxford, UK.

Aubertot, J-N. (2004). The effects of sowing date and nitrogen availability during vegetative stages on Leptosphaeria maculans development on winter oilseed rape. *Crop Protection* 23 (juillet): 635-645. doi:10.1016/j.cropro.2003.11.015.

Aubry C, Papy F, Capillon A (1998). Modelling decision-making processes for annual crop management. *Agricultural Systems* 56:45–65

Barberi P., Lo Cascio B. (2001) Long-term tillage and crop rotation effects on weed seedbank size and composition. *Weed Research* 41 (4) (août 18): 325-340. doi:10.1046/j.1365-3180.2001.00241.x.

Barari H. , Cook SM, Clark S J., et Ingrid H. Williams IH. (2005). Effect of a turnip rape (Brassica rapa) trap crop on stem-mining pests and their parasitoids in winter oilseed rape (Brassica napus) ». *BioControl* 50 (février): 69-86. doi:10.1007/s10526-004-0895-0.

Bartlet E, Blight MM, Pickett JA, Smart LE, Turner G, Woodcock CM (2004). Orientation and feeding responses of the pollen beetle, *Meligethes aeneus*, to candytuft, *Iberis amara. J Chem Ecol* 30: 913–925.

Bianchi, F.J.J.A, C.J.H Booij, et T Tscharntke. (2006). Sustainable pest regulation in agricultural landscapes: a review on landscape composition, biodiversity and natural pest control ». *Proceedings of the Royal Society B: Biological Sciences* 273 (1595) (juillet 22): 1715-1727. doi:10.1098/rspb.2006.3530.

Borg A, Ekbom B (1996). Characteristics of oviposition behaviour of the pollen beetle, *Meligethes aeneus* on four different host plants. *Entomologia Experimentalis et Applicata* 81: 277–284.

Büchi R (1995) Combination of trap plants (*Brassica rapa* var. *silvestris*) and insecticide use to control rape pests. *IOBC/wprs Bulletin* 18(4): 102–121.

Carcamo HA, Dunn R, Dosdall LM, Olfert O (2007). Managing cabbage seedpod weevil in canola using a trap crop - a commercial field scale study in western Canada. *Crop Protection* 26: 1325–1334.

Chatelin, M. H., Aubry, C., Poussin, J. C., Meynard, J. M., Massé, J., Verjux, N., Gate, P., and Le Bris, X. (2005). DéciBlé, a software package for wheat crop management simulation. *Agricultural Systems* 83, 77-99.

Chauvel B, Guillemin JP, Colbach N, Gasquez J (2001) Evaluation of cropping systems for management of herbicide resistant popula- tions of blackgrass (Alopecurus myosuroides Huds). *Crop Protection* 20:127–137. doi:10.1016/S0261-2194(00)00065-X

Cook, S. M, L. E Smart, J. L Martin, D. A Murray, N. P Watts, et I. H Williams. (2006). Exploitation of host plant preferences in pest management strategies for oilseed rape (Brassica napus) ». *Entomologia Experimentalis et Applicata* 119 (3) (juin 1): 221-229. doi:10.1111/j.1570-7458.2006.00419.x.

Cook, S. M., Zeyaur R. Khan, et John A. Pickett. (2007). The Use of Push-Pull Strategies in Integrated Pest Management. *Annual Review of Entomology* 52 (1) (janvier): 375-400. doi:10.1146/annurev.ento.52.110405.091407.

Cook, S. M., Murray D.A., et Williams I.H.. (2004). Do pollen beetles need pollen? The effect of pollen on oviposition, survival, and development of a flower-feeding herbivore ». *Ecological Entomology* 29 (2) (avril 1): 164-173. doi:10.1111/j.0307-6946.2004.00589.x.

Cook, Samantha M, Elspeth Bartlet, Darren A Murray, et Ingrid H Williams. (2002). The role of pollen odour in the attraction of pollen beetles to oilseed rape flowers ». *Entomologia Experimentalis et Applicata* 104 (1) (juillet 1): 43-50. doi:10.1046/j.1570-7458.2002.00989.x.

Daniel C., Dierauer H. & Clerc M.. (2011). The potential of silicate rock dust to control pollen beetles. Bi-annual meeting of IOBC. 4-6 Ocober 2011. Göttingen. *IOBC/wprs Bulletin (in press).*

David C. Jeuffroy M-H., Recous S., Dorsainvil F. (2004). Adaptation and assessment of Azodyn model for managing the nitrogen fertilisation of organic winter wheat. *European Journal of Agronomy* 21, 249-266

De Bruyn, L., J. Scheirs, et R. Verhagen. (2002). Nutrient stress, host plant quality and herbivore performance of a leaf-mining fly on grass ». *Oecologia* 130 (4) (février): 594-599.

Debaeke, Philippe, Nicolas Munier-Jolain, Michel Bertrand, Laurence Guichard, Jean-Marie Nolot, Vincent Faloya, et Patrick Saulas. (2009). Iterative design and evaluation of rule-based cropping systems: methodology and case studies. A review. *Agronomy for Sustainable Development* 29 (mars): 73-86. doi:10.1051/agro:2008050.

Dejoux J.F. (1999). Evaluation agronomique environnementale et économique d'itinéraires techniques du colza d'hiver en semis très précoces, Thèse de Doctorat INA P-G Paris 243p.

Dejoux J.F., Ferré F., Meynard J.M. (1999). Effects of sowing date and nitrogen availability on competitivity of rapeseed against weeds in order to develop new strategies of weed control with reduction of herbicide use, 10th International rapeseed congress, Canberra (Australia), 1999/09/26-29, CD-Rom New horizons for an old crop

Dejoux J.F., Meynard J.M., Reau R., Roche R., Saulas P. (2003). Evaluation of environment friendly crop management systems for winter rapeseed based on very early sowing dates, *Agronomie* 23 (8): 12. doi:10.1051/agro:2003050.

Dosdall LM, Herbut MJ, Cowle NT, Micklich TM (1996). The effect of seeding date and plant density on infestations of root maggots, Delia spp. (Diptera: Anthomyiidae), in canola. Canadian J ournal of Plant Science 76:169–177

Dosdall LM, Stevenson FC (2005). Managing flea beetles (*Phyllotreta* spp.) (Coleoptera: Chrysomelidae) in canola with seeding date, plant density, and seed treatment. *Agronomy Journal* 97: 1570–1578.

Dosdall, L.M., M.G. Dolinski, N.T. Cowle, et P.M. Conway. (1999). The effect of tillage regime, row spacing, and seeding rate on feeding damage by flea beetles, Phyllotreta spp. (Coleoptera: Chrysomelidae), in canola in central Alberta, Canada ». *Crop Protection* 18 (3) (avril): 217-224. doi:10.1016/S0261-2194(99)00019-8.

Ecophyto R&D, 2009. Vers des systèmes de culture économes en produits phytosanitaires. Tome II. Analyse comparative des différents systèmes en grandes cultures. INRA éditeur (France), 218 p.

El-Khoury, W. ; Makkouk, K. Integrated plant disease management in developing countries. (2010). Journal of Plant Pathology Volume: 92 Issue: 4 Supplement Pages: S4.35-S4.42.

Evans KA, Allen Williams LJ (1989). The response of the cabbage seed weevil (*Ceutorhynchus assimilis* Payk.) and the brassica pod midge (*Dasineura brassicae* Winn.) to flower colour and volatiles of oilseed rape. *Aspects of Applied Biology* 23: 347–353.

Evans KA, Allen-Williams LJ (1998). Response of cabbage seed weevil (*Ceutorhynchus assimilis*) to baits of extracted and synthetic host-plant odor. *Journal of Chemical Ecology* 24: 2101–2114.

Feeny P, Paauwe KL, Demong JN (1970). Flea beetle and mustard oils: Host plant specificity of *Phyllotreta cruciferae* and *P. striolata* adults (Coleopera: Chrysomelidae). *Annual Entomology Social America* 63: 832–841.

Feldman S. R., Alzugaray C., Torres P. S.; Lewis, P. (1997). The effect of different tillage systems on the composition of the seedbank. Weed Research (Oxford) 37, 71-76.

Ferré F., Doré T., Dejoux J.F., Meynard J.M., Grandeau G. (2000). Evolution quantitative de la flore adventice dicotylédone au cours du cycle du colza pour différentes dates de semis et niveaux d'azote disponible au semis 11. colloque international sur la biologie des mauvaises herbes, Dijon (France), 06-08 september 2000

Finch S (1978). Volatile plant chemicals and their effect on host plant finding by the cabbage root fly (*Delia brassicae*). *Entomol Exp Appl* 24: 350–359.

Free JB, Williams IH (1978). The response of pollen beetle, *Meligethes aeneus*, and the seed weevil, *Ceutorhynchus assimilis*, to oilseed rape, *Brassica napus*, and other plants. *Journal of Applied Ecology* 15: 761–774.

Gabrielle, B., Denoroy, P., Gosse, G., Justes, E., Andersen, M. (1998). Development and evaluation of a CERES-type model for winter oilseed rape. Field Crops Res. 57, 95–111.

Gabrielle, B., P. Denoroy, G. Gosse, E. Justes, et M.N. Andersen. (1998). Development and evaluation of a CERES-type model for winter oilseed rape ». *Field Crops Research* 57 (1) (mai 8): 95-111. doi:10.1016/S0378-4290(97)00120-2.

Guide simplifié de techniques alternatives de désherbage des cultures. (2008). Chambre d'agriculture de Cote d'or. Available on www.cetiom.fr.

Gurr, G.M., S.D. Wratten, et J.M. Luna. (2003). Multi-function agricultural biodiversity: pest management and other benefits ». *Basic and Applied Ecology* 4 (2): 107-116.

Hansen, Lars Monrad. (2004). Economic damage threshold model for pollen beetles (Meligethes aeneus F.) in spring oilseed Janusauskaite Janusauskaite rape (Brassica napus L.) crops ». *Crop Protection* 23 (1) (janvier): 43-46. doi:10.1016/S0261-2194(03)00167-4.

Hokkanen HMT (1989). Biological and agrotechnical control of the rape blossom beetle *M* Jornsgard *eligethes aeneus* (Coleoptera, Nitidulidae). *Acta Entomol Fenn* 53: 25–29.

Hopkins RJ, Ekbom B (1999). The pollen beetle, *Meligethes aeneus*, changes egg production rate to match host quality. *Oecologia* 120: 274–278.

Janusauskaite, Dalia, et Steponas Ciuberkis. (2010). Effect of different soil tillage and organic fertilizers on winter triticale and spring barley stem base diseases ». *Crop Protection* 29 (8) (août): 802-807. doi:10.1016/j.cropro.2010.04.002.

Jeuffroy M.H., Valantin-Morison M., Champolivier L., Reau R. (2006). Azote, rendement et qualité des graines : mise au point et utilisation du modèle Azodyn-colza pour améliorer les performances du colza vis-à-vis de l'azote, OCL, 13 (6), 388-392.

Jornsgard B., Rasmussen K., Hill J., Christiansen JL. (1996). Influence of nitrogen on competition between cereals and their natural populations. Weed research, 36, 461-470.

Kirkegaard, Ja, Pa Gardner, Jf Angus, et E Koetz. (1994). Effect of Brassica break crops on the growth and yield of wheat ». *Australian Journal of Agricultural Research* 45: 529. doi:10.1071/AR9940529.

Kruidhof, H M, L. Bastiaans, et M J Kropff. (2008). Ecological weed management by cover cropping: effects on weed growth in autumn and weed establishment in spring ». *Weed Research* 48 (6) (décembre 1): 492-502.

doi:10.1111/j.1365-3180.2008.00665.x.

Kurstjens, D A G, et M J Kropff. (2001). The impact of uprooting and soil-covering on the effectiveness of weed harrowing ». *Weed Research* 41 (3) (juin 25): 211-228. doi:10.1046/j.1365-3180.2001.00233.x.

Lançon, Jacques, Jacques Wery, Bruno Rapidel, Moussa Angokaye, Edward Gérardeaux, Christian Gaborel, Dramane Ballo, et Blaise Fadegnon. (2007). An improved methodology for integrated crop management systems ». *Agronomy for Sustainable Development* 27 (juin): 101-110. doi:10.1051/agro:2006037.

Le Gal, P.-Y., A. Merot, C.-H. Moulin, M. Navarrete, et J. Wery. (2010). A modelling framework to support farmers in designing agricultural production systems ». *Environmental Modelling & Software* 25 (2) (février): 258-268. doi:10.1016/j.envsoft.2008.12.013.

Le Gal, P.-Y., P. Dugué, G. Faure, et S. Novak. (2011). How does research address the design of innovative agricultural production systems at the farm level? A review ». *Agricultural Systems* 104 (novembre): 714-728. doi:10.1016/j.agsy.2011.07.007.

Lieven J., Quéré L., Lucas J.-L., (2008). Oilseed rape weed integrated management: concern of mechanical weed control. International Endure congress: diversifying crop protection. 13-15 octobre 2008 La Grande Motte France. www.endure-network.eu.

Loyce, C, J.P Rellier, et J.M Meynard. (2002)a. Management planning for winter wheat with multiple objectives (1): The BETHA system ». *Agricultural Systems* 72 (1) (avril): 9-31. doi:10.1016/S0308-521X(01)00064-6.

Loyce, C., J.P. Rellier, et J.M. Meynard. (2002)b. Management planning for winter wheat with multiple objectives (2): ethanol-wheat production ». *Agricultural Systems* 72 (1) (avril): 33-57. doi:10.1016/S0308-521X(01)00065-8.

Lucas, J. A. (2010). Advances in plant disease and pest management ». *The Journal of Agricultural Science* 149 (S1) (décembre): 91-114. doi:10.1017/S0021859610000997.

Lutman, Bowerman, Palmer, et Whytock. (2000). Prediction of competition between oilseed rape and Stellaria media ». *Weed Research* 40 (3) (juin 1): 255-269. doi:10.1046/j.1365-3180.2000.00182.x.

MédièneS., Valantin-Morison M., Jean-Pierre Sarthou, Stéphane de Tourdonnet Marie Gosme,, Michel Bertrand, Jean Roger-Estrade, Jean-Noël Aubertot, Adrien Rusch, Natacha Motisi, Céline Pelosi, Thierry Doré. Agroecosystem management and biotic interactions. A review. (2011). *Agronomy for Sustainable Development.* DOI 10.1007/s13593-011-0009-1

Meynard JM, Doré T, Lucas P (2003) Agronomic approach: cropping systems and plant diseases. CR Biol 326:37–46. doi:10.1016/ S1631-0691(03)00006-4

Munier-Jolain N, Deytieux V, Guillemin JP, Granger S, Gaba S (2008). Conception et évaluation multi-critères de prototypes de systèmes de culture dans le cadre de la Protection Intégrée contre la flore adventice en grandes cultures. Innov Agron 3:75–88

Nilsson, C. (1994). Pollen Beetle (Meligethes aeneus spp) in oilseed rape crops (Brassica napus L.): Biological interactions and crop losses. Dissertation. Sweden: Swedish University of Agricultural Sciences.

Ould-Sidi, Mohamed-Mahmoud, et Françoise Lescourret. (2011). Model-based design of integrated production systems: a review ». *Agronomy for Sustainable Development* 31 (3) (mars): 571-588. doi:10.1007/s13593-011-0002-8.

Potting RPJ, Perry JN, Powell W (2005) Insect behavioural ecology and other factors affecting the control efficacy of agro-ecosystem diversification strategies. *Ecological Modelling* 182: 199–216.

Price, P.W. 1991. « The Plant Vigor Hypothesis And Herbivore Attack ». *Oikos* 62 (2): 244-251.

Primot, S, M Valantin-Morison, et D Makowski. (2006). Predicting the risk of weed infestation in winter oilseed rape crops . *Weed Research* 46 (février): 22-33. doi:10.1111/j.1365-3180.2006.00489.x.

Rapidel, Bruno, Bouba S. Traoré, Fagaye Sissoko, Jacques Lançon, et Jacques Wery. (2009). Experiment-based prototyping to design and assess cotton management systems in West Africa . *Agronomy for Sustainable Development* 29 (décembre): 545-556. doi:10.1051/agro/2009016.

Rapidel, Bruno, Cécile Defèche, Bouba Traoré, Jacques Lançon, et Jacques Wery. (2006). In-field development of a conceptual crop functioning and management model: A case study on cotton in southern Mali . *European Journal of Agronomy* 24 (4) (mai): 304-315. doi:10.1016/j.eja.2005.10.012.

Rathke, G, T Behrens, et W Diepenbrock. (2006). Integrated nitrogen management strategies to improve seed yield, oil content and nitrogen efficiency of winter oilseed rape (Brassica napus L.): A review. *Agriculture, Ecosystems & Environment* 117, n°. 2 (11): 80-108. doi:10.1016/j.agee.2006.04.006.

Reau R. et Wagner D (1998). Les préconisations du CETIOM. Oléoscope, 48 : 10-13.

Rusch A, Valantin-Morison M (2009) Effect of nitrogen fertilization, cultivar and species on attrac- tiveness and nuisibility of two major pests of winter oilseed rape (*Brassica napus* L.): Pollen beetle (*Meligethes aeneus* F.) and stem weevil (*Ceutorhynchus napi* Gyl.). *IOBC/wprs Bulletin*.

Rusch A., Muriel Valantin-Morison, Jean Roger-Estrade, Jean-Pierre Sarthou· (2011). Effect of crop management and landscape context on insect pest populations and crop damage. *Agriculture, Ecosystems and Environment* (in press)

Rusch, Adrien, Muriel Valantin-Morison, Jean Pierre Sarthou, et Jean Roger-Estrade. (2010a). Integrating Crop and Landscape Management into New Crop Protection Strategies to Enhance Biological Control of Oilseed Rape Insect Pests. Dans *Biocontrol-Based Integrated Management of Oilseed Rape Pests*, éd par. Ingrid H. Williams, 415-448. Dordrecht: Springer Netherlands.
http://www.springerlink.com/index/10.1007/978-90-481-3983-5_17.

Rusch, Adrien, Muriel Valantin-Morison, Jean-Pierre Sarthou, et Jean Roger-Estrade. (2010b). Biological Control of Insect Pests in AgroecosystemsEffects of Crop Management, Farming Systems, and Seminatural Habitats at the Landscape Scale: A Review. Dans *Advances in Agronomy*, 109:219-259. Elsevier.
http://linkinghub.elsevier.com/retrieve/pii/B9780123850409000062.

Schott, C. Mignolet C., Meynard J.-M. (2010). Les oléoprotéagineux dans les systèmes de culture :évolution des assolements et des successions culturales depuis les années 1970 dans le bassin de la Seine OCL, 17 (5) 276-291

Shennan C (2008). Biotic interactions, ecological knowledge and agriculture. Phil Trans R Soc Lond B 363:717–739. doi:10.1098/ rstb.2007.2180

Simon, Sylvaine, Laurent Brun, Johanny Guinaudeau, et Benoît Sauphanor. (2011). Pesticide use in current and innovative apple orchard systems . *Agronomy for Sustainable Development* 31 (février 25): 541-555. doi:10.1007/s13593-011-0003-7.

Tscharntke, T., R. Bommarco, Y. Clough, T.O. Crist, D. Kleijn, T.A. Rand, J.M. Tylianakis, S. van Nouhuys, et S. Vidal. (2007). Conservation biological control and enemy diversity on a landscape scale . *Biological control.*

Valantin-Morison M, Quere L (2006) Effects of turnip rape trap crops on oilseed rape pests in a network of organic farmers' fields. *CD-Rom Proc Int Symp Integrated Pest Management in Oilseed Rape, 3–5 April 2006, Göttingen, Germany.*

Valantin-Morison M., Jeuffroy M.H., Saulas L., Champolivier L. (2003). Azodyn-rape : a simple model for decision support in rapeseed nitrogen fertilisation. 11th International rapeseed congress, Copenhaguen (Denmark), 2003/07/6-10, Towards Enhanced value of cruciferous oilseed crops by optimal production and use of the high quality seed components. Proceedings GCIRC

Valantin-Morison, Muriel, et J. M. Meynard. (2008). Diagnosis of limiting factors of organic oilseed rape yield. A survey of farmers' fields . *Agronomy for Sustainable Development* 28 (décembre): 527-539. doi:10.1051/agro:2008026.

Valantin-Morison M. (2011). How to design and assess integrated crop management methods for winter oilseed rape in a network of farmers' fields ? Bi-annual meeting of IOBC. 4-6 Ocober 2011. Göttingen. *IOBC/wprs Bull (in press)*

Valantin-Morison, M, J Meynard, et T Dore. (2007). Effects of crop management and surrounding field environment on insect incidence in organic winter oilseed rape (Brassica napus L.) . *Crop Protection* 26 (août): 1108-1120. doi:10.1016/j.cropro.2006.10.005.

Van Delden A., Lotz LAP, Bastiaans L., Franke AC, Smid HG, Groeneveld RMW, Kropff MJ. (2002). The influence of nitrogen supply on the ability of wheat and potato to suppress Stellaria media growth and reproduction. *Weed Research,* 42, 429-445.

Van Ittersum M.K., Leffelaar P.A. van Keulen H., Kropff M.J., Bastiaans L., Goudriaan J. (2003). ON approaches and applications of the Wageningen crop models. *European Journal of Agronomy,* 18 : 201-234.

Vasilakoglou, Ioannis, Kico Dhima, Nikitas Karagiannidis, Thomas Gatsis, et Konstantinos Petrotos. (2010). Competitive Ability and Phytotoxic Potential of Four Winter Canola Hybrids as Affected by Nitrogen Supply . *Crop Science* 50: 1011. doi:10.2135/cropsci2009.05.0270.

Vereijken, P. (1997). A methodical way of prototyping integrated and ecological arable farming systems (I/EAFS) in interaction with pilot farms . *European Journal of Agronomy* 7 (septembre): 235-250. doi:10.1016/S1161-0301(97)00039-7.

Vollmann, Johann, Helmut Wagentristl, et Wilfried Hartl. (2010). The effects of simulated weed pressure on early maturity soybeans . *European Journal of Agronomy* 32 (4) (mai): 243-248. doi:10.1016/j.eja.2010.01.001.

Vos J., van der Putten PEL, (1997), Field observations on nitrogen catch crops. I potential and actual growth and nitrogen accumulation in relation to sowing date and crop species. *Plant soil* 195, 299-309.

Whytock, C. (1993) The competitive effects of broad-leaved weeds in winter oilseed rape. Brighton crop protection conference, weeds. Proceedings of an international conference, Brighton, UK, 22-25 November.

Wilby, A., et M.B. Thomas. (2002). Natural enemy diversity and pest control: patterns of pest emergence with agricultural intensification . *Ecology Letters* 5 (3): 353-360.

Williams, I. H. (2010). The Major Insect Pests of Oilseed Rape in Europe and Their Management: An Overview. In: Williams, I. H. (Eds.), Biocontrol-Based Integrated Management Of Oilseed Rape Pests. Springer London, pp. 1-44.

# Permissions

The contributors of this book come from diverse backgrounds, making this book a truly international effort. This book will bring forth new frontiers with its revolutionizing research information and detailed analysis of the nascent developments around the world.

We would like to thank Fabio R. Marin, for lending his expertise to make the book truly unique. He has played a crucial role in the development of this book. Without his invaluable contribution this book wouldn't have been possible. He has made vital efforts to compile up to date information on the varied aspects of this subject to make this book a valuable addition to the collection of many professionals and students.

This book was conceptualized with the vision of imparting up-to-date information and advanced data in this field. To ensure the same, a matchless editorial board was set up. Every individual on the board went through rigorous rounds of assessment to prove their worth. After which they invested a large part of their time researching and compiling the most relevant data for our readers. Conferences and sessions were held from time to time between the editorial board and the contributing authors to present the data in the most comprehensible form. The editorial team has worked tirelessly to provide valuable and valid information to help people across the globe.

Every chapter published in this book has been scrutinized by our experts. Their significance has been extensively debated. The topics covered herein carry significant findings which will fuel the growth of the discipline. They may even be implemented as practical applications or may be referred to as a beginning point for another development. Chapters in this book were first published by InTech; hereby published with permission under the Creative Commons Attribution License or equivalent.

The editorial board has been involved in producing this book since its inception. They have spent rigorous hours researching and exploring the diverse topics which have resulted in the successful publishing of this book. They have passed on their knowledge of decades through this book. To expedite this challenging task, the publisher supported the team at every step. A small team of assistant editors was also appointed to further simplify the editing procedure and attain best results for the readers.

Our editorial team has been hand-picked from every corner of the world. Their multi-ethnicity adds dynamic inputs to the discussions which result in innovative outcomes. These outcomes are then further discussed with the researchers and contributors who give their valuable feedback and opinion regarding the same. The feedback is then collaborated with the researches and they are edited in a comprehensive manner to aid the understanding of the subject.

Apart from the editorial board, the designing team has also invested a significant amount of their time in understanding the subject and creating the most relevant covers. They scrutinized every image to scout for the most suitable representation of the subject and create an appropriate cover for the book.

The publishing team has been involved in this book since its early stages. They were actively engaged in every process, be it collecting the data, connecting with the contributors or procuring relevant information. The team has been an ardent support to the editorial, designing and production team. Their endless efforts to recruit the best for this project, has resulted in the accomplishment of this book. They are a veteran in the field of academics and their pool of knowledge is as vast as their experience in printing. Their expertise and guidance has proved useful at every step. Their uncompromising quality standards have made this book an exceptional effort. Their encouragement from time to time has been an inspiration for everyone.

The publisher and the editorial board hope that this book will prove to be a valuable piece of knowledge for researchers, students, practitioners and scholars across the globe.

# List of Contributors

**Andrew A. Efisue**
Department of Crop and Soil Science, Faculty of Agriculture, University of Port Harcourt, Choba, Port Harcourt, Nigeria

**Marcelo de Almeida Silva**
São Paulo State University / UNESP - College of Agricultural Sciences, Brazil

**Marina Maitto Caputo**
University of São Paulo / USP -"Luiz de Queiroz" College of Agriculture, Brazil

**Jude J. O. Odhiambo**
University of Venda, Thohoyandou, South Africa

**Wayne D. Temple and Arthur A. Bomke**
University of British Columbia Faculty of Land and Food Systems, Vancouver, B.C., Canada

**Lei Dou**
College of Agriculture, Ibaraki University, Ibaraki, Japan
Soil and Fertilizer Institute, Shandong, China

**Masakazu Komatsuzaki**
College of Agriculture, Ibaraki University, Ibaraki, Japan

**Fábio Ricardo Marin**
Scientific Researcher, Embrapa Agriculture Informatics, Campinas, SP, Brazil

**Muriel Valantin-Morison**
UMR 211 d'Agronomie INRA-AgroParisTech Bât. EGER - BP01, F-78850 Thiverval-Grignon, France

**Jean-Marc Meynard**
INRA Département SAD - Bât. EGER - BP01 F-78850 Thiverval-Grignon, France

Printed in the USA
CPSIA information can be obtained
at www.ICGtesting.com
JSHW011327221024
72173JS00003B/80

9 781632 391315